Indonesia's Vast New Oil Province Exploration and Production Sharing Contracts as of End of 1972

PERTAMINA'S OIL & GAS DISCOVERY AREAS SINCE 1966

PRINCIPAL DISCOVERY AREAS BY FOREIGN CONTRACTORS SINCE 1966

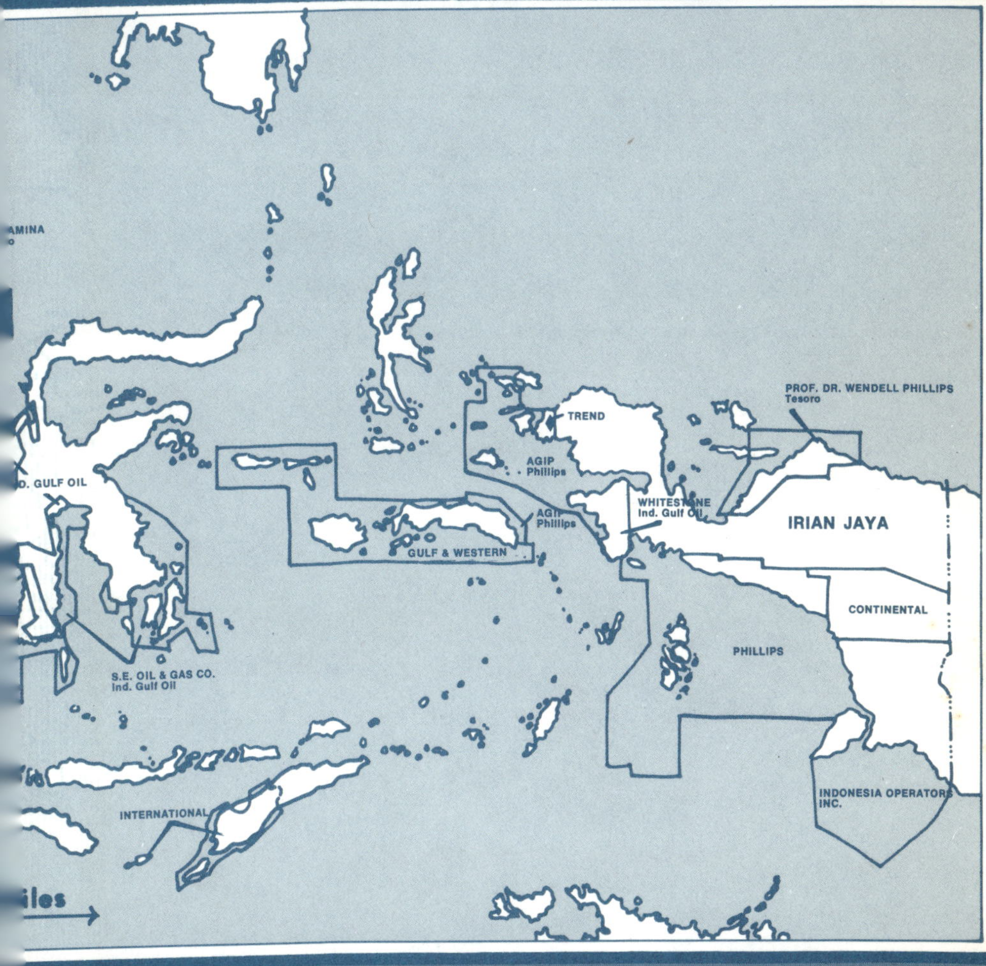

Ruth Sheldon Knowles

Indonesia Today

The Nation That Helps Itself

Nash Publishing, Los Angeles

Library of Congress Catalog Card Number: 73-83530
International Standard Book Number: 0-8402-1325-5

Published simultaneously in the United States and Canada
by Nash Publishing Corporation, 9255 Sunset Boulevard,
Los Angeles, California 90069

Printed in the United States of America

First Printing

I dedicate this book,
with admiration, respect, and love,
to the Indonesian people,
who fought for their freedom and are demonstrating
how wise use of it helps man achieve
his ultimate goals.

Contents

1
Past is Prologue

"Why do your magazines and newspapers write such bad things about me?" President Sukarno of Indonesia asked. "That is not the Sukarno I know!"

It was late 1963. We were sitting in his spacious private office in Merdeka (Freedom) Palace in Jakarta sipping iced vermouth. The handsome, charming, flamboyant leader of the world's fifth most populous country was dressed in a smart gray military uniform with a dazzling array of ribbons and medals, and he wore at a jaunty angle the black velvet Muslim cap that was his trademark. The room was filled with stunning paintings of revolutionary scenes and Indonesian landscapes beautiful Balinese teakwood carvings, and priceless Chinese figurines. Dominating all the works of art, however, was a huge portrait of Sukarno on an easel.

Gesturing toward the portrait, President Sukarno spoke with passion:

> That picture was painted in 1945, on the day after independence. When that Sukarno took over Indonesia we were a hundred million people with twenty-six dialects living in three thousand islands. Now we have one language and we have unity. Only 8 percent of the people knew how to read and write. Now 86 percent know how. When I was a student at the university there were only seventeen Indonesian students. Now there are two hundred thousand. Is that not an achievement? Look what I have done for my people. They are happy and they like to dance. Why do your senators call me corrupt and a bandit? What does Sukarno do? Does he rob people? Does Sukarno kill people? Does Sukarno rape women? Sukarno does not even own a house. It belongs to the people. Yes, I get free light and free water–and free drinks, when I serve my guests like I am serving you. But do you know how much salary Sukarno gets? I only get ten thousand rupiahs a month–not a day–and rupiahs, not dollars. They say that Sukarno is a Communist. Sukarno is *not* a Communist. I am the only world leader without a party behind him, because the people are behind Sukarno.

The word picture of himself which Sukarno painted for me was a masterpiece of surrealism. The Sukarno of 1945 had indeed been the Father of His Country, liberating his people from 270 years of Dutch colonial control and molding them into a nation. For almost two decades he had been idolized and beloved by the people, who affectionately called him Bung (elder brother) Karno. However, the Sukarno of 1963, who had just been elected president for life by a rubber-stamp congress, had blinded himself with grandiose overconfidence to the fact that his "happy" people had the lowest income in Asia, food and clothing were at a bare minimum, inflation was running wild, and the nation was bankrupt.

Rice bowls could not be filled with Sukarno's splendid oratory or glowing promises, but like a magician he created illusions of progress to appeal to the intense national pride that he fostered. He was a master at playing the free world against the Communist world for all the money in aid and loans he could get from both of them and had plunged the country into $2.4 billion of foreign debt. The bulk of it was for the prestige projects and military adventuring he used to sustain patriotism at a high pitch and justify economic scarcity, as well as to satisfy his insatiable need for revolutionary excitement.

The Sukarno of 1963 saw only his successes. He had just hosted the elaborate, costly Asian games, having, for the occasion, given his capital an impressive new look with a multimillion-dollar sports stadium, luxury hotel, magnificent fountains, statues, new buildings, and a divided highway with a cloverleaf intersection. "I must feed my people's souls, not only their stomachs," he explained. He had just forced the Dutch to capitulate to his claim for control of Irian Jaya, formerly Netherlands New Guinea, following a massive Indonesian military buildup on a billion dollars of credit from Russia for supplies.

Although he triumphantly promised the people he would "overcome the bottlenecks and the difficulties of economic problems in a short time," he then recklessly precipitated the nation into a military and economic confrontation with the newly created Federation of Malaysia. (The federation, a result of Great Britain's decolinization program, brought together Malaya and the former British colonies of Singapore, Sarawak, and British North Borneo.) Hailing the federation as a British plot to encircle Indonesia, Sukarno's battle cry was, "Crush Malaysia!" While he called for more sacrifices on the part of the people, he lived lavishly, as did a corrupt group of sycophants around him.

Sukarno was blithely unaware that he had already set in motion forces that would destroy him in two stormy, dramatic years. He had fallen victim to his own charisma. He was a romantic revolutionary in love with the ideology of revolution. Sukarno was constantly "rediscovering" his own revolution and exhorting the people "to keep fanning the flames of the leaping fire of the revolution. Let us become logs to feed the flames of revolution." He told them repeatedly, "I am your lips. I am your voice. I am your mouth. I am Indonesia. I am the revolution."

The sound of his own voice deafened Sukarno to the rumblings of discontent and disillusionment among the people. Nor did he listen to the advice of those brilliant and loyal patriots around him who could see that he was embarked upon a collision course. Instead he let himself be wooed by the Communists. When he declared his "Crush Malaysia" campaign he accepted the advice of Chinese Premier Chou En-lai to form a fifth branch of the Indonesian armed forces under Communist or pro-Communist command.

The Indonesian Communist party was waiting for just such an opportunity to infiltrate the anti-Communist army in large numbers and receive military training and weapons. Under Sukarno, Communist party membership had grown to 3 million, which made it the largest Communist party in the world outside China and the Soviet Union. It was also one of the hardest working, and had gained some 20 million supporters by indoctrinating youth and women's groups, labor unions, and rural workers.

As Sukarno floundered in the quicksands of his confrontation with Malaysia, he found himself more and more within Peking's sphere of influence. Angered by American criticism, he lashed back with "Go to hell with your aid!" When Malaysia was admitted to the United Nations Security Council for a one-year term, he dramatically announced Indonesia's withdrawal from U.N. membership. Sukarno encouraged anti-American demonstrations and harassment of Americans by Communist groups who wrecked U.S. Information Service libraries. Electricity, gas, and mail services to U.S. installations in Jakarta were suspended. American newsmen were forbidden the use of telecommunications facilities. The United States cut off all foreign aid, and a plan to organize an international consortium to aid Indonesia was abandoned. Sukarno had already passed a death sentence on the Indonesian economy by halting the commerce with Malaysia in which almost 30 percent of Indonesia's trade customarily flowed. It was obvious that Indonesia was becoming lost to communism. It was just a question of time.

What happened next caught everyone by surprise—those in both the Western and the Communist worlds and, most of all, Sukarno.

The political volcano began to erupt in Jakarta just before dawn on October 1, 1965. A group of Communist conspirators from the army and the Communist party launched simultaneous raids on the homes of seven of the army's highest ranking generals, including the army's commander in chief, either killing them on the spot and taking their bodies or kidnapping them—with the exception of General Nasution, minister of defense and chief of staff of the armed forces, who managed to escape. The conspirators, who called themselves the September 30th Movement, seized the presidential palace, the Radio Indonesia station, the central telecommunications, and an air base just outside the capital. A few hours after the raids they proclaimed on the radio that a Revolutionary Council of Indonesia had acted to seize control of the country to forestall a counterrevolutionary plot, led by a Council of Generals in the army and sponsored by the American Central Intelligence Agency, to take over the country. President Sukarno was mysteriously incommunicado.

The Communists had made a fatal mistake, which changed the subsequent course of Indonesian history. They had not placed on their abduction list General Suharto, a relatively obscure officer who was chief of the army's Strategic Command. In less than three hours after the raids General Suharto took charge of loyal army forces in the capital. By nightfall he talked the rebel troops into surrendering and retook the palace and the radio and telecommunications stations without a shot being fired. Later, in a statement that typifies his character and methods, he said, "I could have taken these posts by force in the morning, but I thought it senseless to incur the bloodshed we could have then expected. I preferred a 'silent action.' "

General Suharto had learned in the early morning that President Sukarno had gone to the captured air base, where the September 30th Movement had established headquarters. He sent a messenger asking Sukarno for instructions and, when none were forthcoming, sent another advising Sukarno to leave the base because Suharto was preparing an attack. Sukarno left for his summer palace at Bogor, forty miles south of Jakarta. The army and civilian leaders of the September 30th Movement fled by air. General Suharto's task force retook the base without casualties.

When General Suharto and other army commanders conferred with him at Bogor Palace, President Sukarno did not clarify his role nor indicate that he knew anything about the kidnapped generals. Sukarno appointed a weak general to be in charge of army operations, commissioned General Suharto to restore security and order, and broadcast a message to this effect to a bewildered public. Two days elapsed before the mutilated bodies of the kidnapped generals were discovered stuffed in a well near the air base. The grisly work of excavation was covered by television, radio, and newspapers. A horrified public reacted with mass violence. Fighting between anti-Communists and Communists flared up throughout the country. Before the bloody purges subsided and General Suharto restored order, an estimated three hundred thousand persons were killed.

It did not become known until much later that President Sukarno was aware of the plot to eliminate the anti-Communist army leaders and, being persuaded they were a threat to him, went along with the conspirators' plans. Nor did he realize that, as evidence ultimately revealed, the Communists planned to use him as titular head of the nation while they gradually took over. President Sukarno was pointedly absent from the funeral for the murdered generals. He called the assassinations "a ripple in the ocean" of the Indonesian revolution. He was blind and deaf to the new revolution that had begun.

For six chaotic months Indonesia was on the verge of civil war. While President Sukarno ranted about furthering the revolution and protected the Communist party and left-wing cabinet ministers, the people launched fresh waves of Communist killings and demanded that the Communist party be outlawed. What began as anticommunism soon became anti-Sukarnoism under the leadership of university and high school students. They demanded that General Suharto oust Sukarno, reorganize the government, ban the Communists, and improve the economy.

The climax came when President Sukarno arrogantly dismissed General Nasution as minister of defense, swore in a new left-wing cabinet, banned the student organizations, and closed the University of Indonesia. General Suharto and the conservative army leadership were reluctant to challenge President Sukarno directly, but demonstrations and new revolutionary fever in the capital created such an emergency that they had no choice.

General Suharto was ill, but on March 11, 1966, he ordered three leading generals to see President Sukarno and convince him that the only way to restore calm and civil order was to transfer the presidential executive powers to Suharto. Generals Basuki Rachmat, M. Yusuf, and A. Machmud had a stormy, all-day session with Sukarno. He finally yielded and signed a statement transferring his powers to Suharto but retained his titles of president, supreme commander, and great leader of the revolution. General Suharto's first official act the following day was to outlaw the Communist party—in the president's name. The capital exploded in a great victory celebration.

With the rejection of Sukarno it was obvious that Indonesians were giving themselves a second chance. But had they run out of time? The nation lay devastated, in economic and emotional chaos. Even the most optimistic and sympathetic observers believed in March 1966 that Indonesia faced a perilous future and many years would be required to begin to develop its potential. Instead, in a miraculously short time, Indonesia has given the modern world its most spectacular example of how a nation can be reborn by helping itself. Sukarno had sidetracked the Indonesian revolution, but the idealism and determination of the generation of 1945 that had achieved it were still intact in the leaders who now took his place. Furthermore, like most Indonesians—Sukarno being an exception—they were pragmatists. They had to look no farther than their own country to see what was wrong with socialism and communism, and they were determined to rewrite the book. The new leaders had the insight to know that a free-enterprise system would rejuvenate the country and had the courage to make all the right economic decisions without worry about the political consequences. They have solved one of the greatest problems plaguing the less-developed countries: how to reconcile nationalism with the encouragement of foreign investment to develop their rich natural resources and import technology and new skills.

Indonesia reached a landmark year in 1973. President Suharto was given another five-year term by a new consultative congress elected by the people. This was the first time they had gone to the polls in sixteen years; and it also marked the beginning of the last year of President Suharto's first Five Year Development Plan. Under it, his New Order set an unprecedented record for a developing country by exceeding targets rather than failing to meet them. Inflation was only 2½ percent in 1973 as contrasted to 650 percent in 1966. There has been a 2,300 percent increase in domestic savings since 1968. There is a trade surplus. Through its new policies, Pertamina, the national oil company, has obtained $1.5 billion of private foreign oil exploration and development capital, thereby opening up a vast new oil province, doubling oil production, and making Indonesia the world's eighth largest oil-exporting nation. Since the promulgation of the 1967 Foreign Capital Investment Law, the government has approved 549 development projects, outside of oil; these represent a commitment of $2.4 billion of private foreign investment, of which 20 percent is by American firms. Because of its resources, both human and natural, Indonesia has emerged as the most important developing country in the Asian Pacific region open to foreign investment. It is a country whose boom times are just beginning. As a businessman in Jakarta remarked to me, "Ten years from now everyone who isn't here will be kicking himself for not having come to Indonesia in the early 1970s." Indonesian economists are now leaders in international monetary affairs. Finally, due to the strategic location of the three-thousand-mile-long archipelago at a major crossroads of the world—between the Pacific and Indian oceans and between the continents of Asia and Australia—and due to the way it is developing its vast natural resources, Indonesia is assuming a stabilizing political and economic role in Southeast Asia and becoming a major world power.

On a recent trip to Indonesia I visited President Suharto in his personal office. Both the man and the setting were in startling contrast to my visit with President Sukarno. Suharto is a sturdy man of fifty-two. He is known as "the smiling general," but his is not a facing-the-camera smile. It signifies his attitude. It illuminates his face with warmth and understanding. It reflects the new quality in the Indonesian people today: self-confidence. Suharto was dressed in smartly tailored, gray slacks and a tropical tunic open at the throat. Over the heart was the colorful coat of arms of the Republic of Indonesia. I had noticed that all his cabinet ministers wore this attire in a variety of colors. I asked Suharto if he had designed it as an official "uniform." He smiled and said, "I have not given specific instructions that my ministers should wear it, but I thought I would set the example that this is a perfect costume to wear working in your office in a hot climate and is distinctive enough to receive ambassadors formally. Also, the insignia will remind every minister of his responsibility and duty to his country."

The modern building that houses his executive offices was designed by President Suharto to serve also as headquarters for the national development campaign. Adjoining his private office are spacious galleries filled with beautifully displayed exhibits including models, charts, and maps that detail every phase of Indonesian development activities and are kept up to date. The galleries provide a workshop in which the cabinet can visualize progress. They are also open to public and student tours for the study of the constantly changing story of the new Indonesia.

President Suharto in the simple attire, with the coat of arms of the Republic of Indonesia, which he and his cabinet members wear to symbolize the spirit of the New Order.

Businessmen and diplomats find the exhibits extremely useful; for example, on one occasion the Egyptian ambassador and his staff spent two days studying them to learn about development.

As President Suharto discussed with me the progress that had been made and the problems ahead, he spoke always of the people. I mentioned that the change I noticed most in Indonesia was the expression on their faces. They were cheerful, smiling, and animated, which was not often the case in Sukarno's time. "I have been visiting villages throughout the country to see for myself the change in the people," President Suharto said. "They have caught the spirit of development." It is this spirit, from the president down to the simplest farmer in a rice paddy, that is the yeast in Indonesia's loaf of bread which continues to rise.

In numerous trips to Indonesia since 1966 I have had the opportunity to witness the amazing transformation of this ancient and beautiful country and to know the remarkable, dedicated men who have been responsible for it. The story of how they have launched a new era of democracy and increasing prosperity is highly significant in today's world in view of United Nations studies showing that the gap between the rich and the poor nations is steadily widening. There is growing pessimism about the capacity of less developed countries to solve their problems. Indonesia also has many thorny problems still unsolved, but its record of problem solving to date and its forward planning leave little doubt that it will achieve its goals. Indonesia today offers an exciting blueprint that works and which can show other countries how they, too, can build soundly for their future.

2
First Steps

The oldest and most popular form of dramatic art in Indonesia is the *wayang* shadow-puppet play that tells the ancient myths and legends of Indonesia as well as the religious Hindu epics from the *Ramayana* and the *Mahabharata.* The stories focus on the eternal struggle between good and evil, in which good is always triumphant. They are a major form of entertainment for young and old alike, at all levels of society. Wayang performances begin early in the evening and last until daybreak. The tales are long and complicated with a multitude of characters and the audiences are enthralled and patient. Throughout Indonesia's history the wayang has been one of the principal ways in which values and ethics are taught. It has been important in teaching new ideas and beliefs and during the revolution it was used to tell the people about the aims of the struggle for freedom.

In March 1966, when President Sukarno was persuaded to sign over his powers to General Suharto, a complicated wayang

An Indonesian puppeteer with his painted puppets, made from buffalo hide, which he uses in the "wayang" shadow plays, Indonesia's most popular dramatic art, that tell the ancient legends of the eternal struggle between good and evil.

drama began which was difficult for the outside world to understand because of its peculiarly Indonesian implications and techniques. Obviously a country cannot function successfully with two leaders who have entirely different goals. However, in Indonesia there is rarely such a thing as direct confrontation accompanied by immediate action. The people may run amok, as they did in the wave of anti-Communist killings, but historically the Indonesian way of achieving change is by *musyawarah*—arriving at a consensus through deliberations. It was this philosophy that underlay a period of difficult dualism with President Sukarno trying to recapture his power and General Suharto, acting in the president's name, moving slowly but steadily to turn power over to the people constitutionally and legally.

The experience and character of General Suharto, the career soldier, were the antithesis of those of the volatile, political Sukarno. Suharto, the son of a landless Javanese peasant, started his military career in colonial days when at the age of nineteen he was accepted in the Royal Netherlands Indies Army military school. The Dutch army offered one of the few opportunities for a peasant boy to find any sort of career. During Japanese occupation in World War II he was a company leader in the volunteer Indonesian defense army trained by the Japanese. It was disbanded at the end of the war in the Pacific but immediately reorganized as a revolutionary army to fight the returning Dutch. Suharto, now a lieutenant colonel, was an outstanding military leader in the ensuing three years of bloody struggle for independence. He did not concern himself with the seething maelstrom of politics surrounding Sukarno. His contacts with the president were infrequent, and their relationship was that between loyal officer and commander in chief.

Although Suharto in 1966 was in fact commander in chief, he was insistent that his powers be confirmed by the Supreme People's Consultative Congress, which he reconvened. Most of its members had been appointed by Sukarno, but when he imperiously refused to give the congress a report on his role in the abortive coup, the congress made it plain that Sukarno's days of grandstanding were over. The congress not only confirmed his emergency transfer of power to General Suharto but reduced President Sukarno's legal title of great leader of the revolution to an honorary status and revoked his presidency for life, although it permitted him to remain as president. General Nasution, whom Sukarno hated violently, was unanimously elected chairman of the congress, making him second in importance to Suharto, inasmuch as the office of vice-president remained vacant. General Nasution, who was the army chief of staff and had planned and led the guerrilla warfare against the Dutch, was a dedicated nationalist and strong anti-Communist. His philosophy was that the army was the guardian of the constitutional government.

In the events precipitating the March 1966 crisis, the demands of the student organizations focused on three objectives: outlawing the Communist party, ridding the cabinet of pro-Communist members, and lowering prices. Following his ban of the Communist party Suharto arrested eighteen pro-Communist and pro-Sukarno ministers and formed a new cabinet to tackle economic problems and unify the armed forces as well as the other political parties. Two key civilian cabinet members were Adam Malik, foreign minister, and Sultan Hamengku Buwono IX of Jogyakarta, minister of economics, finance, and development.

Adam Malik had a remarkable background: spirited journalist, revolutionary, politician, and diplomat. After serving as ambassador to the Soviet Union and Poland he told friends, "Communism is not the path that Indonesia should follow." He had worked consistently and outspokenly within the government to counteract growing Communist power, and in the process he had become anti-Sukarno as well as anti-Communist.

The sultan of Jogyakarta was one of the most respected and influential leaders in Indonesia. A descendant of Javanese nobility, he was a strong nationalist and the only sultan to support the revolution actively. He was educated in Holland and studied law at the University of Leiden. He held many cabinet posts under Sukarno, who regarded him as a potential rival, but the sultan had never joined a political party. He was the revered spiritual and temporal leader of some two million people in the traditional sultanate that now had the status of a province of which he was governor.

Digging into the government's appalling economic mess, Suharto and the cabinet realized they must act drastically on all fronts at once to keep the economy from total collapse. Only an emergency infusion of foreign aid from non-Communist countries could help. Sukarno had effectively alienated every potential source, so the most urgent task was to demonstrate that Indonesia was reversing its policies and would be a good credit risk. Malik began negotiations to end the ridiculous three-year confrontation with Malaysia and indicated that Indonesia would like to rejoin the United Nations. The United States responded with an $8-million credit for rice, the British with a $2.8-million gift, and the Japanese with a $30-million loan for imports. It was a beginning.

General Suharto genuinely did not want to see Sukarno publicly discredited, but the old buffalo continued to paw the earth and snort, making politically provocative speeches and maneuvering behind the scenes. In an astonishing display of unrealistic rhetoric, Sukarno told the nation in his Independence Day speech of August 1966 that he was still prime minister and supreme commander of the armed forces and that Suharto had no real power. He declared he had agreed to the peace treaty with Malaysia only after it had been greatly changed and that Indonesia would not immediately recognize Malaysia nor rejoin the United Nations. He attacked the United States and railed against imperialism. He thundered to the crowd, "Follow my leadership, obey all my directives. March onward to continue the revolution along the path of my direction!" Students booed him, and most of the huge crowd was hostilely silent.

The final act of Suharto's and Sukarno's personal wayang began with the trials of Sukarno's ministers. As the full story of their Communist connections, intrigues, and personal corruption running into millions upon millions of dollars unfolded, the

Indonesian people were shocked. President Sukarno's refusal to testify soon placed him in the position of seemingly being tried by proxy. For the first time, an uncensored Indonesian press printed sensational stories of Sukarno's political adventuring and charged that although he had four wives as permitted under Muslim law, love affairs were his hobby. All this was common international knowledge but unknown to the majority of Indonesians. Sukarno became a figure of public ridicule and contempt. The people's mood became ugly when Sukarno's complicity in the assassination of the generals became apparent. Thousands of students demonstrated, crying, "Hang him! Hang him!" They demanded that "Citizen Engineer Sukarno" be tried and sentenced.

Public tension mounted to such a peak that in March 1967 General Nasution convened an emergency session of the congress which appointed General Suharto acting president until July 1968 when popular elections would be held, revoked all President Sukarno's constitutional powers of government, and prohibited him from engaging in any political activities until the general elections.

This was an Indonesian solution. President Sukarno still retained his title, and the congress specifically left any further actions against him to the discretion of the new acting president. In his acceptance speech Suharto tried to save face for the Father of Independence. He told the people that the congressional decision not to oust Sukarno but to take away his political authority was "wise and correct" and "based on psychological considerations in society." He referred to him sympathetically as Bung Karno and told them a team of doctors had confirmed that Bung Karno's health had deteriorated so that it was "more than proper that we should treat him according to the condition of his health." He reminded them that "President Sukarno's position" had been confirmed by the nation's highest authority, the congress, and said, "So let all of us no longer discuss the problem of the position of Bung Karno."

Suharto's tact was wasted on Sukarno, who continued to hold forth at the palace as though he were still in control.

Regretfully, Acting President Suharto decreed that Sukarno be stripped of all titles, prohibited from flying the presidential flag, and forced to reside at the summer palace. However, he was allowed to wear his uniform, ribbons, and medals "when attending state occasions at the official invitation of the government." The personal wayang had ended except for an epilogue that took place on the first Independence Day after Suharto became acting president. Palace officials preparing for the ceremony could not find the key to the box containing the original red-and-white national flag raised on the day Indonesia proclaimed independence, which had been traditionally raised every Independence Day thereafter. Sukarno had taken the key with him. He gave it up, but when the box was opened the flag was missing. There were indignant public demands that he be forced to return the flag. However, Suharto diplomatically invited Sukarno to give the flag to the nation in a formal ceremony. Sukarno's compliance symbolically marked the solid establishment of the New Order; and he passed into obscurity and died three years later.

As long as the colorful Sukarno was a political factor, the real character of the self-effacing Suharto was greatly overshadowed and misinterpreted both at home and abroad. He was neither a "military dictator" nor a "political compromiser," as he was variously protrayed. He was what he had always been—a conscientious, honest, competent, patriotic, religiously devout, career military officer who had spiritually never left the village in which he was born. These were essential qualities for leadership of the staggering reconstruction job the nation faced. In addition, President Suharto had that rarest of qualities—a clear mind, with the ability to see to the heart of the most complicated problem and logically reach a solution. He was a superb logician, the best in the army and the best in the country.

Now that the period of dualism was ended Suharto could move more directly to achieve what he had pinpointed from the beginning as a chief cause of the nation's troubles. On the same day he had outlawed the Communist party he had issued a second announcement that, being less dramatic, had not yet entered into the mainstream of public consciousness. He urged all national entrepreneurs to improve production, distribution, and services, saying, "Only through the activities of the entrepreneurs can the sufferings of the people be averted, the people from whom we were born and whom we serve. Following this, security and public order will materialize." This was strange language in a country that since independence had been indoctrinated to believe in state controls and expropriation and nationalization of private property as the means to cure economic woes.

Suharto had never read an economics textbook, but he had the practical economic values of his peasant upbringing in a family where there was scarcely enough money to pay the five-cents-a-month tuition when he began school. However, poverty alone had not been the best economics teacher. Indonesian army officers had to provide food, housing, and supplies for their soldiers, so for twenty years Suharto had been an entrepreneur himself in organizing and directing agricultural and industrial projects and cooperatives and in training his soldiers in the art of self-reliance as well as soldiering.

Suharto was determined to make self-reliance a national goal. From the beginning he spoke plainly to the people. "I'll not promise there will be all-out changes overnight, that the living of the people will all of a sudden change to the better," he told them. "Such a promise will be cheating the people and cheating ourselves. To establish economic stability and rehabilitation is an extremely difficult task. This is the task of the whole people."

There was no way immediately to meet the demand of the students and public for the lowering of prices, but for the first time they had real hope that it would be done. They now had a government that acted rather than talked. When President Suharto did make a speech, it was to tell them in simple terms exactly what the state of the country's economy was and what the government was doing about it.

Indonesia's plight was evident. In 1966 foreign currency reserves amounted to only $8 million, but the country's debts were 300 times this amount, $2.4 billion. Interest and repayment charges on the debts amounted to $675 million that year whereas the nation's export earnings had sunk to around $360 million. Most of this had to go for emergency imports. Industry had ground almost to a standstill. Consumer prices had increased 10,000 times since 1951. The people had not yet felt the full impact of what was, in effect, a total collapse, because the purchasing power of the urban public had been partially protected by subsidies and because the other 70 percent, rural, population could live by subsistence agriculture.

Agricultural products, rubber, coffee, copra, and pepper, together with forestry and fishing accounted for about 70 percent of the value of Indonesia's exports, with petroleum accounting for most of the rest. Export taxes on agricultural products produced nearly 20 percent of the government's revenues. The most important agricultural cash crop was rubber, which yielded about a third of the total value of exports. The big rubber estates, which produced about 60 percent of the total rubber crop, had been nationalized and had deterioriated to the point where over half of them were not financially viable. Almost all of the small holders of rubber trees needed to replant, but with world rubber prices declining and the cost of living rising they lacked money to stay in business. Rice was the most important component of the cost of living, and rice allocations were a significant part of wages paid to the military, civil servants, and the employees of government agricultural estates and industries. Yet rice production had not kept up with population growth. There was a shortage of fertilizer, irrigation systems had silted due to neglect, and distribution systems had broken down. Beginning in 1961 more than a million tons of rice were imported annually. Because of a shortage in exchange, imports in 1965 were cut to 140,000 tons, creating acute shortages and soaring prices. Inflation peaked in 1966, when it rose 635 percent.

President Suharto and his government team waged their economic war with the same fervor and dedication with which they had fought for independence. Adam Malik and the sultan of Jogyakarta won a major battle to reestablish international confidence as they flew from capital to capital in the non-Communist world, explaining the New Order's plans and policies. Indonesia rejoined the United Nations with a flourish, as a nonaligned nation. Beginning in September 1966 conferences were held between Indonesia and her main Western creditors—the United States, West Germany, France, Italy, Great Britain, the Netherlands, and Japan—to work out a plan to consolidate and reschedule the payments of Indonesia's foreign debts. These conferences, also attended by representatives from Canada, Australia, Switzerland, and New Zealand, were supported by the International Monetary Fund and the World Bank. They resulted in the formation of the International Governmental Group on Indonesia, or IGGI, which rescheduled the entire sum of debts, including interest falling due up to the end of 1968, and gave Indonesia a three-year grace period before payments would begin. The IGGI also agreed to a fresh loan in 1967 of $200 million on soft terms of 3 percent annual interest, twenty-five years to repay, and a grace period of seven years. The United States, which pledged separately to meet a third of this amount if the other donors met the rest, gave even softer terms: forty years to repay, a ten-year grace period, 1 percent annual interest during the grace period, and 2.5 percent thereafter. In 1968 the IGGI loaned another $400 million in a similar arrangement and promised more loans to come. The International Monetary Fund established a permanent mission in Jakarta to assist the Indonesian Central Bank and the Ministry of Finance. The World Bank established a major permanent mission, the first one in a less-developed country, to provide technical assistance on project planning and implementation.

Indonesia's chief creditor was the Soviet Union which loaned $990 million, representing more than 40 percent of the total debt. Relations with the Soviets were understandably cool, but Adam Malik went to Moscow and in a series of hard-bargaining conferences persuaded them to reschedule this debt.

At the same time the government was busy negotiating all this essential aid, it was strengthening its regional cooperation in Southeast Asia in order to lay a broader base for increased trade. Indonesia played an important role in setting up in 1967 the Association of Southeast Asian Nations (ASEAN), for economic, social, and cultural cooperation, which grouped together Indonesia, Malaysia, the Philippines, Thailand, and Singapore. By also establishing trade contacts with Taiwan, Indonesia further froze her relations with Communist China, which had been suspended following the abortive coup.

The greatest proof to the international community of Suharto's 180-degree turn to the right came in December 1966 when he decreed the return of all properties, plantations, and plants which had formerly been owned by foreign investors and nationalized by the government. A billion dollars worth of Dutch assets that had been seized in 1957 were negotiated under a separate agreement by which Indonesia agreed to pay $167 million in compensation. Operations were resumed by most of the foreign owners including such companies as Unilever, British-American Tobacco, Bata, Heineken, Nicholas Aspro, Prodenta, Singer, National Cash Register, Union Carbide, and Goodyear.

A month later the welcome mat was laid out for new foreign investors when the congress passed the 1967 Foreign Capital Investment Law permitting direct private foreign investment in all sectors except defense industries and public utilities. Tax holidays, easy repatriation of profits and capital, special advantages for foreign investment in outlying islands, and two-year free-import privileges for raw materials, machinery, and the personal needs of foreigners—these and other incentives were offered to attract foreign capital and know-how into joint or completely foreign-owned ventures. Businessmen from the United States, Western Europe, Japan, and neighboring Asian countries began to crowd Jakarta's Hotel Indonesia to investigate investment and trade opportunities.

James Linen, chairman of the executive committee of *Time* magazine, attended the opening of a new resort hotel in Bali which had been started by Sukarno and Intercontinental Hotel Corporation, a subsidiary of Pan American Airways. Linen was so impressed with what the New Order was doing that he asked the sultan of Jogyakarta what he could do to help. The sultan suggested he could make international business aware of Indonesia's new opportunities. Linen organized a conference in Geneva, sponsored by Time, Inc., which was attended by the heads of leading business corporations of all kinds from the United States and Europe. The sultan headed a team of ten Indonesian specialists in all principal fields to make presentations. Adam Malik accompanied the team to discuss the political situation. "I insisted that they have a number of panel discussions," the sultan told me at the time, "because we expected to learn as much from the conference as those who attended. We had passed our investment laws, but the proper implementation of them is of vital importance. We need to know from businessmen what is necessary for them to be able to do business."

In addition to restoring foreign investors' confidence in Indonesia, President Suharto had the equally difficult task of restoring Indonesians' confidence in foreign investors. "At the beginning, of course, there existed an uneasiness about the role of capitalism in the country," the president told me, explaining:

> Some people adhered to the theory that if foreign capital comes to Indonesia it is the beginning of capitalism, which means the beginning of imperialism. But I tell them that they do not need to be afraid, as we know how to regulate it in the appropriate way so that the flow of capital will serve to benefit the people of Indonesia and the government and also the investors of the countries concerned. The investment of capital in Indonesia must be done in such a way that besides accelerating the exploitation of Indonesian natural resources, the Indonesian people must be provided with skills and technology so that the time will come when we will be able to explore and exploit our natural resources by ourselves.

On the home front, President Suharto had his priorities clearly in mind. He considered economic and political stabilization inseparable. Only if he could improve the people's standard of living, especially in regard to food and clothing, could the political situation be stabilized so that general elections could be held to restore true constitutional government and achieve democracy. Obtaining an economic breathing spell from the enormous burden of debt and reestablishing credit through a new, independent foreign policy were the necessary first steps toward Suharto's goal of national self-reliance. However, success depended on finding an economic program that would work—quickly!

Early in 1966 Suharto called on the keenest economic brains in Indonesia to chart the nation's specific course. This advisory team, which became a formal one in 1968 and many members of which became ministers, was comprised mainly of liberal, free-market economists from the University of Indonesia, most of whom had doctorates from American universities. This task force was directly responsible to the president. The members were all ardent patriots who had chafed under the ruination of their country. The advisory team was also a powerful committee to handle day-to-day economic decisions to achieve economic stability. Most of the team had links with the army: before 1965 they had often lectured to the army staff at the Command School and taught army seminars dealing with military involvement in political, economic, and social affairs. They had helped shape the army's thinking, thereby contributing to its strong anti-Communist stand.

Suharto's economic team brought a remarkable variety of talents to the government. One of the members was Sumitro Djojohadikusumo, a brilliant economist with a doctorate from the Netherlands School of Economics. He had held several ministries under Sukarno, and as dean and professor of economics at the University of Indonesia he had trained most of the economists who were now Suharto's brain trust. In 1958, having sided with an armed rebellion against Sukarno which failed, Sumitro went into voluntary exile, stating he would not return as long as Sukarno was in power. He became a renowned international economic and financial consultant to government institutions and international corporations and in 1967 returned to Indonesia eager to help rebuild the country. When Suharto formed his Development Cabinet in 1968 he made Sumitro minister of trade.

Among the other important economists who helped make the day-to-day decisions was Widjojo Nitisastro, who became chairman of the National Development Planning Council. Widjojo had a doctorate from the University of California at Berkeley and when he joined Suharto's team was dean of the economics faculty at the University of Indonesia and simultaneously chief of the National Economic and Social Research Center. Widjojo was the leader of the Indonesian delegation to creditor country meetings on aid to explain the New Order's economic policies and problems.

Emil Salim, another Berkeley Ph.D., also was a member of the delegation; he was deputy chairman of the new National Development Planning Council as well as a member of the Provisional People's Consultative Congress.

Ali Wardhana, who became minister of finance in the Development Cabinet, was one of Sumitro's students who had obtained his doctorate from the University of California at Berkeley. He taught at the University of Indonesia and was an economic advisor to the governor of the Central Bank.

Mohammed Sadli, who had a degree in civil engineering from Gadjah Mada University and one in economics from the University of Indonesia, received his economics doctorate from Massachusetts Institute of Technology and had been a fellow at Harvard University's Center of International Affairs. President Suharto appointed him minister of manpower and chairman of the Technical Committee on Capital Investments.

Leaders of other countries, including the United States, also have assembled economic advisory teams to cope with emergencies, but too often sound economic advice plays second fiddle to political expediency. Not so with General Suharto.

"From the very beginning, whenever we discussed solutions to problems, President Suharto would ask only one question, 'What is best for the country?' "Sadli told me.

> When President Suharto first asked us to help, we didn't think it would last very long, but he wanted us to act as well as advise. We all wear three hats—we are the economic team, we are in the cabinet, and we still teach at the university. One of the important things about President Suharto is that he has never had an inferiority complex. We all had to lose our superiority complexes. His decision-making process is without equal. He wants a full report together with alternative solutions. Then he makes up his mind without delay.

To stabilize the economy, the New Order included strong steps. Its financial austerity program for fighting the hyperinflation was built on the classic fiscal and monetary measures of trying to balance the government's budget and restraining the issuance of bank credits. Of some three hundred prestige projects started by Sukarno, such as new buildings and an unrealistic steel mill, over half were put into mothballs. Import duties were raised sharply and domestic tax collection was intensified. Expansion of government employment was curtailed, but government employees were given a 50 percent increase in salaries and 25 percent increase in rice rations. Interest rates on public savings were increased to mobilize development investment funds.

President Suharto was also mobilizing the efforts of the people. Only 10 percent of the country's fifty thousand miles of roads were in even fair condition, which accounted for agricultural surpluses in some areas and shortages in others. The farmer couldn't get to market. Suharto put a large part of the Indonesian army to work repairing roads.

The Ministry of Manpower achieved remarkable results speedily by establishing "food for work" projects using U.S. surplus food commodities aid. A typical project was Demak in central Java, an extensive coastal plain inhabited by half a million people. During the rainy season the rivers crossing the plain were flooded. During the dry season the rivers sank, leaving the surrounding areas parched, cracked, and unproductive. Alternating flood and drought had left the people destitute, and thousands of them migrated to adjoining cities to become beggars and vagrants. Those who remained suffered from malnutrition and were forced to dig the bottoms of dried-up canals in search of drinking water. The manpower ministry organized some eight thousand needy workers to remove more than a million cubic meters of sludge and dirt from the ancient canal system, which had been neglected since the departure of the Dutch. They were paid for their work with four pounds of food daily for a family of five, in the form of bulgur wheat, corn, oil, and a high-protein food blended of corn, soy beans, and milk. Total cost of the food provided by the United States was around a half a million dollars, but dollars and cents is an inadequate measure of the value of the economic benefits of this project. After six months of work, the water that began flowing in the rehabilitated canals gave more than forty thousand people a new start in life. It assured them of two crops of rice every year and the opportunity to grow corn, cassava, vegetables, and fruits. The villages, filled with new hope, had such pride of achievement in the undertaking that they voluntarily constructed secondary and tertiary canals that brought water to small land-holdings throughout the formerly devastated area. This was a demonstration of how to develop self-reliance. The success of the Demak project encouraged the manpower ministry to carry out similar projects throughout the island of Java.

Although there was a rice crisis in 1967, caused by draught and inadequate government stockpiling and procurement of foreign supplies, the New Order managed to weather the storm of public protests against spiraling prices and to draw new battlelines on stabilization of rice prices and increase of production. Since rice makes up about 40 percent of the average Indonesian's market-basket purchases, government policies literally concentrated on and began at the rice roots of the nation. During this crisis President Suharto began what have become known as his "silent inspection" visits to see for himself what was happening in his country. Dressed in civilian clothes, he made unannounced visits to rice markets, mills, and paddies and talked with traders and farmers. Finding malpractices and illegal speculators, he took action against them. The importance of these inspections, however, was the confidence engendered in the people by seeing their president directly involved in their most pressing problem.

Under Sukarno corruption had become a way of life, from the highest to the lowest civil servant. During the trials of Sukarno's ministers innumerable scandals of corrupt practices were exposed. One businessman testified that he paid Sukarno $600,000 for a special deferred-payment license to import yarn, worth about $10 million. Small bribes were an established way of obtaining equipment and services. At the ports, customs and military officials were notorious for having their hands out for payoffs to deliver goods, which otherwise would be delayed. At banks, businessmen would be asked for commissions of 10 percent for the grant of a loan or charged "private interest rates" on top of the official State Bank rates. The most serious form of corruption was the widespread smuggling of imports and exports. This was also the most difficult to control because of Indonesia's island geography and interisland trade.

President Suharto and his team realized that corruption was not only a major political problem but a serious economic one that could thwart much of their economic planning and action. As a first step, Suharto organized a strong anticorruption team, headed by the attorney general, with an advisory board consisting of the minister of justice and the commanders of the army, navy, air force, and police. They began an impressive clean-up program. Port administration was removed from army control, and the army was no longer permitted to import most goods tax-free. Following investigations, numerous officials and employees of the Customs Bureau, State Revenue Office, Pension Bureau, and State Banks were dismissed. One of the government's chief foreign exchange earners was a state-owned rubber corporation located in Atjeh in north Sumatra. Seven officials, including the director, were arrested and charged with malpractices that were responsible for the "collapse" of the enterprise.

President Suharto knew the fight against corruption would be a long, continuing one. Opportunities and incentives for corrupt practices could be dried up only by increasing the flow of goods, reducing export taxes on key items, eliminating red tape, raising salaries, and similar measures. He told the people, "I will throw a net over those who are being corrupt and start pulling it tight at the corners. In order to escape it they will have to become leaner and leaner in order to get through the holes, instead of becoming fatter and fatter."

The word *economics* comes from the Greek word *oikonomika,* meaning "skilled in the management of a household." No developing country ever had such a thorough house-cleaning as did Indonesia during 1967 and 1968 in preparation for the big economic push yet to come–the Five Year Development Plan initiated in April 1969. The primary goals of political and economic stabilization had been achieved.

The various political parties had not been able to agree on an election law for the 1968 election, at which time Suharto's appointment as acting president expired. Consequently, the Provisional People's Consultative Congress elected Suharto president until 1973 and provided for popular elections of a new congress in 1971. Under the Indonesian Constitution, the congress elects the president and vice-president. The postponement of popular elections gave political parties time to organize. It also gave the government time to launch its development program.

The Five Year Plan was immensely practical, based on all the guidelines President Suharto and his team had already put into action. It set long-range targets in every sector but wisely provided that the plan be implemented on the basis of detailed annual plans. It gave priority to agriculture and all agriculture-

Indonesia's economy begins at the rice roots. This rice paddy is a typical scene in Java.

related activities in order to give the farmer tools, fertilizer, improved seeds, credit, and incentive to increase production as rapidly as possible. The Five Year Plan also emphasized rehabilitation and extension of the infrastructure, which would speed up the flow of goods and communication. There would be "rice roads" from production to consumption centers and "export roads" to move crops to ports. In the industrial sector, priority was given to those industries which produce chemicals, machinery, and equipment for agricultural use and to those which use such products of agriculture as food, beverages, tobacco, and textiles, timber products, and paper.

Next to agriculture, the Five Year Plan considered the development of Indonesia's great petroleum and mineral potential to be of the most importance. Recognizing that this involves high-risk enterprises that require specialized technology and enormous capital outlays, the plan stressed the necessity of inviting the participation of foreign capital.

The government reaffirmed throughout every section of the Five Year Plan that its policy was to stimulate private capital investments, domestic as well as foreign, and that its role was that of guidance and direction in order that the private sector could develop "soundly and properly." The plan stated that the government would engage only in development undertakings such as infrastructure and strategic projects that were not profitable or attractive to private investment. This was the government's wide-open-door policy for growth, expressing its expectation and belief that the country would reap a rich harvest that could not be estimated but was bound to come.

On the pragmatic financial side, the Five Year Plan gave a leading place to the inflow of official foreign capital aid from the IGGI countries, projecting that 57 percent of the total investment funds for the plan would come from them. However, with justifiable pride the government announced that one-fifth of the first year's development budget would come from government savings. In 1968 the policy of a balanced budget had made it possible to use all foreign credits for financing development, whereas routine expenditures were met entirely by domestic revenues.

To explain the meaning of the plan to the people and their part in it, President Suharto enlisted the help of thousands of professional and amateur puppeteers who added a new economic story to the old tales for village entertainments.

Indonesia was demonstrating its determination to implement development on the basis of its own strength and ability. Furthermore, events were already under way that would increase its capacity to do so beyond the most optimistic hopes of President Suharto and his economic team.

3
"Tanah Air Kita"

Geography has had a profound effect in shaping the history, culture, racial composition, and economic development of Indonesia. Its political fate has always been bound up in its geographic location, its physical resources, and the conflicts between those who have tried to reduce those resources to possessions. Geographic environment, in its broad aspect, is the study of men and their relationships and interactions in the development of the physical resources they find at their disposal as well as their ability to trade for or seize the resources of others to satisfy either needs or avarice. Thus, Indonesia's success today is based on its geography and natural resources.

The Indonesian archipelago of more than thirteen thousand islands, of which about three thousand are inhabited, stretches west to east along the equator between the Indian and Pacific oceans for thirty-two hundred miles, a distance equivalent to that from San Francisco to Bermuda. From north to south the archipelago extends eleven hundred miles between Asia and

Australia. The total land and sea area is nearly four million square miles, a million square miles more than the land area of the contiguous United States. Indonesia's sea area is about four times larger than its land area, and Indonesians have always called their country "Tanah Air Kita"–Our Land and Water. It shares with Malaysia command of the Strait of Malacca, which is the shortest sea route from the Suez and India to China, Japan, and North America. Between its islands pass alternative routes from Australia north to the Orient and North America. Consequently, Indonesia has historically been involved in migrations from Asia, world commerce, and the conflicts of colonial powers.

More than 90 percent of the country's land area is in five big islands. The smallest of these is Java, about the size of the state of New York, but more than half of Indonesia's 120 million people live on this fertile, intensively cultivated island. Its population density of 1,442 persons per square mile is one of the highest in the world. The other big islands–Kalimantan (the southern part of Borneo), Sumatra, Sulawesi, and Irian Jaya (the western part of New Guinea)–are sparsely populated and, although rich in natural resources, only now beginning to be developed. Many primitive Stone Age tribes still live in the mountainous rain forest jungles of both Kalimantan and Irian Jaya.

The archipelago is one of the world's most volcanic areas, and volcanic cones, many of them active, form the impressive chain of peaks which march majestically through the chain of islands. The earth's crust is still in formation in this region. Minor eruptions and earthquakes are daily occurrences, but the benefits from volcanoes far outweigh the harm. The matter and ash ejected from volcanoes weather into extremely fertile soil. The forest-covered volcanic rocks collect and store rainfall that becomes available for agriculture through surface streams. Since the equator bisects the major islands of Indonesia, there is abundant, continuous rainfall and no winter or summer. In the past hundred years Jakarta's meterological station, the oldest in the tropics, has never recorded temperatures higher than 96 degrees or lower than 66 degrees.

All these favorable physical and climatic conditions combine to provide the archipelago with a rich tropical agricultural potential. The heavy rainfall makes its flora the most varied in the world with more than thirty thousand species of trees, shrubs, and grasses belonging to the more than twenty-five hundred families already recorded. The earth is rich in minerals: petroleum, tin, nickel, coal, iron, copper, lead, bauxite, gold, silver, and diamonds.

One of Indonesia's greatest resources which has had a profound effect upon its people is the incredible beauty of its diverse geographic environment. Travel through the islands is an experience of never-ending delight and contrasts–sparkling mountain lakes, streams, and rivers; green jungles abounding with tigers, elephants, and rhinoceroses; grassy plains, terraced rice paddies, and coral beaches. Everywhere is an abundance of exotic flowers, ferns, birds, and butterflies. All this beauty has become part of the Indonesian people's character, religion, literature, drama, music, and art. It is reflected in their highly artistic dances, costumes, textiles, woodcarvings, and silverwork. The Indonesians have held a mirror up to nature.

The Republic of Indonesia's motto is "Unity In Diversity," a phrase used to describe the country by a fifteenth-century Indonesian poet. Few other nations have developed from such diverse influences and variety of peoples. Although primitive people inhabited the Indonesian archipelago more than half a million years ago, when ice sheets still covered Europe and North America, the islands were settled by migrations from Asia that began about 3000 B.C. The people are mainly of Malay stock but include some 140 groups differentiated in type, religion, and language.

During the first century of the Christian era Hindu traders from India found a well-developed culture of organized communities of rice farmers and fishermen who traded with the Asian mainland. The Hindus established settlements and were assimilated by the Indonesians, who adopted Hindu culture and its various forms of religion, including Mahayana Buddhism. By the sixth century there were many rich Indonesian-Hindu kingdoms in Java and Sumatra, renowned for navigation, commerce, arts, and learning. Indonesia's Golden Age was during the fourteenth century when a Javanese king united all the archipelago into one powerful empire, the kingdom of Majapahit. The unification and achievements of this period, which lasted 126 years, marked the birth of Indonesia and provided a heritage that enabled it to be unified again after the colonial period.

The Majapahit kingdom fell under the impact of the Islamic religion that was introduced by Moslem traders and spread throughout the islands. Many of the Indonesian kings were converted and established strong, rival sultanates. When the Portuguese, Spanish, Dutch, and English began their colonial expansion in the sixteenth century they encouraged the rivalries between Indonesian rulers in their own European power struggle for control of the spice trade. Through the ruthless tactics of the Dutch East India Company, which began trading and waging continuous warfare in 1602, the Dutch won control of a large part of Indonesia by 1680 and treated it as a colony.

The Dutch East India Company reduced the sultanates to economic vassal states through monopoly of trade, the sultans becoming their agents. The once-independent farmers were made serfs by the introduction of a system of economic bondage. The company would advance money at planting time to purchase future harvests at low prices and, when a farmer fell into debt, would confiscate his property. To collect taxes the company brought in Chinese, who added the role of usurious moneylenders and eventually dominated Indonesian internal commerce. When the company was liquidated in 1799 and the Dutch government took over direct and formal control of Indonesia as a colony, the vicious feudal system was perpetuated.

For five brief years during the early nineteenth century the luckless Indonesian peasants saw some hope for a better future. When Napoleon's army conquered Holland, the British, still at war with France, seized Dutch colonies. The British wanted to prevent the lucrative Indonesian exports from reaching Napoleon and also prevent him from encircling India. One of the most colorful Europeans in the history of the Orient, Sir Stamford Raffles, was made British lieutenant general of Java. The Dutch, never interested in the welfare, history, or culture of the Indonesians, viewed them only as economic pawns in their colonial empire. Nor was the British government interested in the Indonesians; the British considered their occupation to be temporary, pending the ultimate outcome of the Napoleonic Wars. However, Raffles, who was a cultured idealist as well as a brilliant commercial agent for the British East India Company, saw his assignment as a crusade to liberate Indonesia from Dutch oppression. He wanted to persuade the Indonesian princes and sultans to invest the British governor-general of India with the title and ancient rights of Majapahit's kings in order that British rule, which he considered just, could be established over the archipelago with the support of Indonesian rulers.

Raffles set about making a complete transformation of the Javanese system of government through a series of legal and judicial reforms. His purpose was to replace Dutch feudalism and indirect rule through the aristocracy as vassals of the Dutch with a modern organization and direct rule of the people by paid government officials. His major reform was to eliminate the Dutch system of forced labor and deliveries of products by setting up a new land-rent system. Taxes, which were levied on villages rather than individuals, varied according to the productivity of the land and could be paid in money or rice. The purpose was to provide economic incentive to the farmer and freedom of choice as to what he would raise. Five years was too short a time for one man to change the whole system. When the Napoleonic Wars ended and Britain returned Java to Holland, the Dutch reinstituted most of their old oppressive practices. However, Raffles, who went on to found Singapore, had added to the Indonesian heritage another chapter that strengthened their pride and resolve to regain their freedom.

In his two-volume *History of Java,* Raffles told the story of a high Dutch official who gave the sultan of Jogyakarta a huge tiger with long white hair. One of the sultan's favorite sports was to stage fights between tigers and buffaloes. When he pitted one of his buffaloes against the gift tiger, which had a reputation for winning many fights, the sultan's buffalo won. The Dutch official was mystified by the unusual excitement this caused in the whole court. He did not realize that in these fights the Javanese compared the tiger to the Dutch and the buffalo to the Javanese. A sultan's buffalo killing an unconquerable Dutch tiger was an omen of great consequence.

The insatiable European appetite since the Middle Ages for the taste of cloves, nutmeg, and pepper caused Indonesia's downfall. Cloves and nutmeg were grown only in its islands. The policy of the Dutch East India Company was to create a monopoly and achieve high prices through scarcity. When the company assumed control it eradicated cloves on all islands except two, Amboina and Ternate. In the latter half of the eighteenth century the French succeeded in smuggling cloves to Indian Ocean islands and to the New World, thereby breaking the Dutch monopoly.

There are few if any apologists for the practices of the Dutch East India Company. Nevertheless, it laid the basis for the realization of Indonesia's agricultural potential when in 1711 it distributed coffee plants among the chiefs in two districts of Java to see if a new commercial crop could be introduced. At that time the world's limited supply of coffee came principally from Yemen in Arabia. The first crop of 100 pounds was gratifying proof of the success of the experiment, but to the company's monopolistic horror, after twelve years the crop from Java's rich volcanic soil reached 12 million pounds, and sultan and peasant alike were profiting. The company set up rigid controls of production to maintain its monopoly and its policy of not permitting Indonesians to profit, but it had set a precedent for agricultural development.

After the company's liquidation, the Dutch government carried on all its activities and exploited Indonesia as though it were one huge government plantation, with compulsory labor, forced deliveries of products, and a monopoly of exports. At the same time it also began to introduce other new possible export crops; these did not bring any immediate results but were to be of long-range benefit to Indonesia's agricultural development. The Dutch experimented with tea seeds and plants from China for almost fifty years with no success. Then in 1873 a tea plant from Assam, a state in northeast India, was introduced; it flourished so in Indonesia that before World War II Java and Sumatra produced 18 percent of world tea exports.

Nineteenth-century production of quinine was limited to South America, where the cinchona tree, from whose bark the medicine was made, is native. The price of quinine was so high that in 1852 the colonial government sent an official to South America to bring back seventy-five cinchona trees. It took twenty years to develop cinchona plantations in Java, but beginning in the twentieth century and until the discovery of a synthetic substitute during World War II, Java produced 90 percent of the world's quinine.

Other introductions included the cassava plant and oil palms. Cassava had been introduced into Indonesia from America by the Spaniards or Portuguese, but it was of poor quality. To give the farmers a second crop on the rice fields, which would enable the government to keep up with the growing population, it brought in a new type of cassava which the farmers accepted. The oil palm was brought from the west coast of Africa, and palm oil products eventually became important exports.

Although freedom was still almost a century away, a new era in Indonesia began when the Liberal party gained control in Holland in the 1860s. Slavery and compulsory labor were abolished. To permit private agricultural enterprises by Indonesians and to encourage the establishment of new ones, monopoly of trade was ended and an agrarian law to formulate rights of ownership to landed property was passed. The law recognized the special character of Indonesian property rights, prohibited the selling to non-Indonesians of land owned or used by Indonesians, and claimed as government property all lands to which no Indonesian held a claim. The government hoped that a system based on free-enterprise principles would increase production, and that income from customs duties and taxes would be greater than the returns from government monopoly operations had been.

Other than by the development of tin mines on two small islands, no interest had been expressed in what Indonesia's subsoil might contain. The discovery of the world's first commercial oil near some oil seeps in Pennsylvania in 1859 launched a worldwide search for this urgently needed new source of illumination, fuel, and lubrication. However, more than a decade elapsed before it occurred to anyone that the numerous oil seeps in Java and Sumatra were signposts to the country's greatest natural resource, both on land and under its waters, which would eventually play the most important environmental role in the development of its human resources.

For centuries Indonesians had found many uses for the oil from these seeps as medicine, torches, caulking for boats, and even as war weapons. Ancient records of the eighth century relate the use of "fire-balls" in naval warfare off the Sumatran coast, and during the sixteenth century the Sumatrans defeated a Portuguese armada by this means. In the early days of the Dutch East India Company, shipments of "earth oil" were sent to Holland, where it was valued as a liniment.

In 1871 a government storekeeper, Jan Reerink, believing that what had happened in Pennsylvania could happen in Indonesia, managed to promote funds from a group in Holland to make the try. He brought a Pennsylvanian drilling rig to west Java, where there were many surface seeps. His great gamble failed: after four dry holes he ran out of money.

No one else was inspired to take up the search until nine years later. In north Sumatra, A.J. Zijlker, superintendent of a Dutch tobacco-planting company, was making the rounds of his company's plantation when a sudden monsoon rain forced him and his Indonesian foreman to take shelter in a tobacco barn. When it became dark, the foreman lighted a torch that flamed more brilliantly than any Zijlker had seen. He asked the foreman if it was made of some exotic resinous wood unknown to Europeans. The foreman explained that the torch had been made in the usual way except that it had been rubbed with "earth wax," a substance obtained by skimming nearby pools of water. He said that earth wax had been used for torches as far back as anyone in that part of Sumatra could remember. The next day, when Zijlker went with the foreman to one of the pools he immediately recognized the familiar odor of kerosene, which was reaching Indonesia from America in great quantities.

Zijlker took a sample to Batavia, as Jakarta was called in colonial times, and a chemist assured him it would make fine lamp oil. Knowing of the phenomenal success of John D. Rockefeller's Standard Oil Company in selling lamp oil to the world, Zijlker went to the sultan of Langkat who ruled over the area near the earth wax springs and obtained a lease from him to explore for and produce petroleum. He formed a company and managed to persuade two Batavian banking houses to advance funds against stock for exploratory drilling. The first well was a dry hole, but the second struck oil in 1885 at seventy-two feet, about a yard deeper than America's first discovery well. It was not a big well, as it yielded only slightly more than five barrels a day, but it was big enough to launch an Indonesian oil boom and to be the cornerstone upon which one of the world's biggest international oil companies, Royal Dutch-Shell, was built.

Other entrepreneurs began to drill wherever they could find oil seeps, and within fifteen years eighteen different companies were producing oil in south Sumatra, east Java, east Borneo, and northeast Borneo. Zijlker's Royal Dutch Company, with backing from Holland, was the biggest and most aggressive. It built Indonesia's first refinery and was anxious to expand and enter the world market. Its largest competitor was Shell Transport and Trading Company, an English concern operating in east Borneo. Shell had been established by Marcus Samuel in London to trade with the Orient. From a modest beginning by importing polished sea shells, which were popular with English ladies who wished to decorate screens, boxes, and knickknacks, he became a major Far East trader in all commodities. Shell, having established a shipping line to transport and sell Russian kerosene in the Orient, had come to Indonesia in search of a more secure supply closer to its markets and, upon finding oil, had also built a refinery.

Standard Oil dominated the markets of the Far East, and neither Royal Dutch nor Shell alone could hope to survive a struggle with Standard to capture its markets there or elsewhere in the world. Together, however, they might be able to surpass it, so after some years of difficult negotiations the two amalgamated to form Royal Dutch-Shell, in which company the Dutch held 60 percent and the English 40 percent. By 1911 the new firm had acquired the rest of the independent Indonesian companies. The following year the powerful young company, which was already marketing in Asia, Europe, Australia, and parts of Africa, invaded the American market by delivering more than a million gallons of Sumatran gasoline to Seattle and distributing it in brand-new Shell railroad tank cars.

The stage was set for a worldwide power struggle between giants which would change the shape of the international oil business for years to come. Furthermore, Indonesia's oil resources became the subject of an international dispute and diplomatic war between the governments of the United States and Holland.

The all-powerful Standard Oil Trust had begun to worry seriously about competition from Indonesian oil several years before the Royal Dutch-Shell merger. At first the trust tried to use the technique upon which its monopoly in America had been so successfully built: the purchase of developed or partly developed properties. It had tried to buy into both Royal Dutch and Shell and been rejected, since neither company wanted to surrender control to the Americans. Standard's efforts to buy other independent companies were thwarted by Royal Dutch-Shell, which had acquired all of them. Royal Dutch-Shell's exports to Asia and Australia now equalled exports to those areas from the United States.

American public concern over the Standard Oil monopoly had resulted in 1911 in the Supreme Court order that

Rockefeller's trust be broken up into thirty eight companies. Of these the largest was Standard Oil of New Jersey, which had been the holding company and continued to be the largest oil company in the world. Since it could not buy into Indonesia, Jersey Standard decided to take the plunge into exploration. Because exploration and mining rights in Indonesia could be held only by Dutch citizens, inhabitants of the Netherlands, or Indonesia, or by a company organized in one country or the other, Jersey Standard, through a Dutch affiliate, organized, in 1912, Nederlandsche Koloniale Petroleum Maatschappij, known as NKPM. The new company got a toehold by obtaining almost a hundred unpromising, scattered concessions most of which had been given up by Royal Dutch-Shell. But NKPM was after bigger game. In south Sumatra the province of Jambi contained many oil seeps and was considered exceptionally promising. Royal Dutch-Shell also wanted the Jambi area and opposed Jersey Standard's efforts to gain concessions there. The Dutch colonial administration was at an impasse. It did not wish to oppose Royal Dutch-Shell, nor did it wish, by giving it the concession, to be attacked for furthering a monopoly that was becoming unpopular in Holland because of its large British participation. The administration saved itself by suspending the Mining Law, as of May 1913, under which concessions were granted.

While waiting for the stalemate to be resolved, Jersey Standard began to explore the concessions it controlled. After six years and the expenditure of $6 million it had drilled sixty-six wells; thirty-nine were dry holes and the others produced a total of only 135 barrels daily as contrasted to Royal Dutch-Shell's 34,000 barrels daily.

Fear of an oil shortage gripped the world in 1917 when America entered World War I. The use of oil had revolutionized warfare as drastically as had the invention of gunpowder. This was a war of fast ships, planes, tanks, trucks, tractors, cars, and ambulances. At the end of the war Lord Curzon, British secretary of state for foreign affairs, proclaimed, "The Allies floated to victory on a sea of oil," pointing out that 90 percent of Allied petroleum needs were supplied by America. The end of the war brought a tremendous increase in demand for oil, since it marked the beginning of the commercial aviation industry, the use of oil to power all the world's ships, and the great expansion of the automotive industry. At the same time, American fields had been drained to the point that fears of an oil shortage were intensified. It was vital that new sources of oil be found.

In Indonesia, in 1921, Jersey Standard pressed its request to the colonial administration for a concession for half of the Jambi area. Its application was refused, and, instead, the whole area as well as other areas in Sumatra and Borneo were granted to a new company, jointly owned by Royal Dutch-Shell and the Dutch colonial government. The U.S. State Department took a strong stand in support of Jersey Standard's claim of discrimination and made representations to The Hague. In the United States, Royal Dutch-Shell was building a strong, fully integrated U.S. subsidiary that was active in every producing area. It applied for leases on federal and Indian lands, but these applications were refused by the U.S. government. The diplomatic war did not affect the outcome of the control of the Jambi area, but it did open the door for Jersey Standard to obtain concessions on 625,000 acres of promising new land in Indonesia and pave the way for the entry of other American oil companies. Royal Dutch-Shell's monopoly had been broken.

Jersey Standard's persistence was to be rewarded in a way that made victory taste particularly sweet. For several years it had been drilling unsuccessfully about eighty miles from Palembang in south Sumatra, on one of the concessions discarded by Royal Dutch-Shell. In mid-December 1921 a well being drilled there had reached the depth at which drilling was generally stopped, so the drilling chief sent a message to the head office as he had been told to do. He kept on drilling, waiting for an answer, but because of the Christmas holidays the order to stop drilling was delayed. On December 26 he struck oil, a few hundred feet below any previously known producing horizon in Indonesia. The Talang Akar discovery well was by no means spectacular, but after a half-dozen more were drilled, one of which produced an initial 2,000 barrels a day, the company realized that it had found a rich field, justifying a refinery. As the field was developed it turned out to be Indonesia's most prolific field prior to World War II.

At the time Jersey Standard received its new concession in 1925, Standard of California, a disaffiliated member of the old trust, had geologists in Indonesia seeking to determine the prospects of competing with its former parent. For ten years it experienced the same treatment of rejection of applications which Jersey Standard had endured, but it, too, was rewarded by persistence. In 1936 it was given a concession in a wild, jungle area in central Sumatra which Jersey Standard had refused. The colonial administration thought it was throwing leftovers from the table to appease the American appetite. Neither it nor Standard of California realized that this remote area contained one of the world's greatest oil fields. The same year Standard of California received the concession it prudently hedged its bet and strengthened its international position by entering into an overseas partnership with Texaco, one of the biggest and most powerful international oil companies. It was an auspicious marriage inasmuch as each had what the other needed, and the union was named Caltex. Texaco had established strong markets around the world but was in need of production to supply them. Standard of California had developed substantial production in Bahrein in the Arabian Gulf and had a concession to explore in Saudi Arabia, but it had no markets.

Caltex of Indonesia began the arduous task of geologically determining where the best spots were to drill and began to plan the logistics of conquering the formidable jungle. Oil seeps, hunches, and trends were no longer the only tools oil seekers had to guide them. A new science—petroleum geology—had evolved since 1910, geologists having discovered from the wells that were drilled that certain rocks always lay in a definite order, each layer revealing a chapter of the earth's history. There were definite clues on the earth's surface which could

Exploration in the dense jungles of Sumatra where no roads exist, such as this Caltex location, is now being carried on by helicopter.

P.T. Caltex Pacific Indonesia

indicate where the rock strata far below arched, dipped, or inclined in opposite directions from a ridge, like an A-frame, indicating that there was a trap, or anticline, below which could have been a favorable place for oil to accumulate. They had also learned that even though the surface gave no indications they could find traps by correlating the fossil forms of plant and animal life found in the various rock strata from wells drilled over a large area. The United States during the 1930s was enjoying its greatest oil boom through the development of the new science of petroleum seismography, a technique of discovering hidden oil structures by generating sound waves with the explosion of dynamite near the surface and recording the travel time to and from buried rock beds in order to determine if they were arched or folded underground and could possibly be oil traps.

Caltex drilled its first prospect in Sumatra in mid-1939 and found indications of oil. A second well found oil in another area. But while the company was preparing to drill its most promising prospect, ominous political events were building up the stormclouds of war which would engulf the world in the tragedy of its second world war.

Forces were also building up inside Indonesia. While the foreigners, both Dutch and American, were absorbed in developing Indonesia's soil and subsoil, Indonesians were developing a nationalistic movement to take the destiny of their land and their waters into their own hands. At the beginning of the twentieth century the Dutch government had announced a new "ethical policy" toward its Indonesian colony, stating that it should be governed not for the sake of the Netherlands but for that of the Indonesians. There was much discussion of self-government and arguments over what kind. The Dutch began to build schools for Indonesians. By the 1920s there were many politically oriented Indonesian organizations—religious groups, study clubs, women's organizations. An Indonesian Communist party had gained strength and in 1926 and 1927 launched armed revolts that the Dutch army crushed mercilessly. The Dutch were frightened by these abortive revolutions, and from then on their policy toward nationalist parties was sternly reactionary. This served only to crystalize the Indonesian goal of total independence rather than self-government.

The fiery, articulate young Sukarno became the leader of a new Indonesian Nationalist party, the strongest political group in the country. After a year in prison for political activities, Sukarno was more active than ever, and the Dutch exiled him, without a trial, to the island of Flores. Sukarno, in his nine years of exile, was the symbol of independence, and his dynamic hold over the people grew rather than lessened in his absence.

When the Japanese invaded Indonesia in 1942 the fleeing Dutch tried to evacuate Sukarno to Australia, but in the confusion he escaped. As the Dutch feared, the Japanese made Sukarno the head of a puppet government, giving him the opportunity to unite his 84 million people during the occupation to prepare for the freedom they were convinced would come.

4
From a Heap of Scrap Iron

The final struggle for freedom was long, bloody, and dramatic. On August 17, 1945, two days after the Japanese surrender, the Proclamation of Indonesian Independence was made with Sukarno as president of the new Republic of Indonesia. Japanese commanders had been ordered to maintain order and security until the arrival of the Allied Forces. A national army was hastily formed, and fighting broke out all over the country for control of the cities. When British and Australian troops with a few small Dutch units, landed two months later, their officers announced that they "would maintain law and order until the time that the lawful government of the Netherlands East Indies is once again functioning." There was a savage battle between British troops and Indonesians at Surabaya in east Java.

As British troops departed they were replaced by Dutch troops, and there was continuous fighting for four years while the Dutch and Indonesians carried on acrimonious discussions.

The Dutch had no intention of giving up their colony and the Indonesians had no intention of giving up their freedom. The Dutch controlled most of the big cities in Java and Sumatra and large areas of the outer islands. The Indonesians held Jogyakarta as their capital in central Java, and their guerrilla armies operated everywhere.

In March 1947 an agreement was signed between Indonesia and the Netherlands, with the good offices of Great Britain, recognizing the de facto sovereignty of the republic over Java, Madura, and Sumatra. Four months after the signing of the agreement Dutch troops invaded all these areas in what they called a "police action," on the flimsy excuse that the Indonesian government had not supplied rice to the Dutch-administered parts of Java. The Dutch then held all deepwater ports and so many principal cities that if the United Nations had not intervened the fledgling republic would have collapsed. A second agreement provided for a military truce and an unstable, unworkable compromise that gave sovereignty to the Dutch until transfer was made to an independent federal United States of Indonesia, in which the new republic would be a minority partner and have control over only the one third of Java it then held.

The Dutch did not intend to keep this agreement, either. In December 1948 they initiated a second "police action," this time to take over all of Java. In a surprise attack with bomber planes and armored tanks they captured the republic's capital and arrested Sukarno and his ministers. The Indonesian army was forced into temporary hiding in the mountains and jungles. There was a world outcry against the Dutch. With traditional stubbornness, they were impervious to all protests and United Nations resolutions until the United States advised them that there would be no Marshall Plan aid for the Netherlands until the Indonesian problem was settled. The Dutch, who were in a desperate financial condition, knuckled under. There was a cease-fire, and the Indonesian leaders were released. An agreement was signed in Jakarta which led to a Round Table Conference in The Hague; there agreement was reached on the

The Pangkalan Brandan memorial monument in north Sumatra of the ruins of Indonesia's first refinery. It is symbolic of how Indonesia's successful state-oil company started with only a heap of scrap iron.

formal transfer of sovereignty over all the former Netherlands East Indies, with the exception of West New Guinea, to the new Republic of Indonesia in December 1949, thus ending 270 years of colonialism.

At Pangkalan Brandan in north Sumatra there is a remarkable monument that symbolizes the ebb and flow of the Indonesian struggle. It is a heap of scrap iron—the remains of the first refinery in Indonesia. It was first destroyed by the Dutch in their "scorched earth" retreat from the Japanese invasion. Patched together by the Indonesians under the Japanese, it was bombed by the Allies. Rebuilt once more by the Indonesians after their 1945 Proclamation of Independence, it was burnt down by them to prevent the Dutch from seizing it in their first "police action." There is a modest stone fence around the rusted jumble of metal, and a memorial plaque states, "No reason exists to be overproud, as this ruin proves."

The monument has an even deeper meaning. The young republic had a long, difficult way to go to learn how to build a nation and develop its resources. This particular heap of scrap iron was important because it represented the physical equipment Permina, Indonesia's national oil company, had when it began the operations that would eventually become Indonesia's greatest economic and nationalistic success.

Occupation and revolution had taken a severe toll of Indonesia's oil industry which, by 1941, had become a principal earner of foreign exchange with its production of 147,000 barrels a day. Anticipating the Japanese invasion,the Dutch had worked out a plan to destroy all oil facilities, since Indonesian oil—the only large source in Asia—was a primary objective, to fuel the Japanese war machine. Refineries were to be burnt, wells plugged with cement, and pipelines dynamited. The Japanese attack was so swift that the Dutch army and the oil companies were not able to destroy all facilities. In south Sumatra, where Japanese troops first landed, the refinery belonging to Stanvac (as Jersey Standard was called following a merger of its Indonesian properties with Socony Vacuum) was totally destroyed, but Royal Dutch-Shell's refinery nearby was still operable.

The Japanese gave priority to rehabilitating oil fields and refineries. They rounded up all of the Indonesian oil-field workers and put them in charge of making repairs. Under colonial practices very few Indonesians held administrative or technical positions, as there were no training schools and top management was exclusively foreign. About the highest position an Indonesian could hope for was that of field foreman after he had learned on the job. The Japanese knew very little about the oil industry, but the few well-trained Indonesians managed to get the industry going again. Production was cut in half during the first year of occupation, but by the next year it was almost up to pre-occupation level. During the last year of occupation production was reduced to 20,000 barrels a day by heavy Allied bombing.

The occupation years were of great importance to Indonesians: they learned that despite Japanese oppression they could run the oil industry themselves. The Japanese set up two schools for workers, and when the managers returned they found a new spirit of self-confidence to match the political independence. This would have a far-reaching effect on the republic's development.

The three big oil companies, Shell, Caltex, and Stanvac, returned to Indonesia shortly after the return of the Dutch military. Despite four years of revolution, by the time sovereignty was transferred to Indonesia their ruined properties were almost completely rehabilitated. Production, which had sunk to 7,000 barrels a day in 1946, was back up to 109,000 barrels in 1949 and within two years exceeded prewar levels.

Caltex was the last to return to its wild, jungle concession area in central Sumatra, but it was an exciting homecoming. Caltex had discovered a little oil by 1942 and had moved a drilling rig to a promising new prospect, but drilling was delayed by the Japanese invasion of Sumatra. The Japanese proceeded to drill the well and at 2,107 feet struck oil at the rate of 800 barrels a day. Ironically, the oil-thirsty Japanese had discovered the Minas field, one of the twenty-two fields in the world to produce over a billion barrels, a mark reached in 1969. Caltex began development of its bonanza, and after its first shipments of Minas oil began in 1952 Indonesia's production began to climb steadily.

Now that Indonesia had control of its land and waters, the problem of how to develop its natural resources for the greatest benefit of Indonesians would be a source of conflict between the government and the oil companies for fifteen years. Under the Round Table Agreement Indonesia had guaranteed to respect the rights of private foreign investors, but the government wanted a greater participation in oil profits because it did not think it was being given a fair share under existing concession agreements. During the war Venezuela had pioneered a new oil law that resulted in a 50-50 profit split between it and the international oil companies. This profit-sharing principle had spread to the Middle-Eastern oil countries. In 1951 the Indonesian parliament suspended grants of new concession or exploration permits until a new mining law could be drafted in accord with changing world conditions.

It would be nine years before an oil law was enacted. A principal reason for the delay was the difficulty of determining the proper role of foreign capital in the nation's development. The state's ultimate goal was to control and develop all its own resources, but Indonesia did not have the capital or skills to do so successfully. Since all new development was suspended, any realization of the country's great potential was put into a deep freeze, which added to its economic woes.

Parliament, also in 1951, gave a thorny task to a Commission of Mining Affairs, to advise whether the government should return to Shell its former concessions in north Sumatra. The north Sumatran concessions were where Indonesia's first oil was struck, forming the basis of the Royal Dutch-Shell combine. In addition to being a shambles this old producing area was embroiled in political activities that made it dangerous and unattractive for Shell to reenter. The possible gain for the company was outweighed by the political risk, which might affect its other properties under the ambiguities of independence. Atjeh, Sumatra's most northern province, was the home of Darul Islam, fanatical Muslims who wanted to establish Indonesia as a Muslim state. After independence they continued guerilla fighting against the new republic. Return of the fields became a heated political issue both locally and in the distant capital in Java. Numerous commissions investigated, and while the matter dragged on for eight years nothing was done to rehabilitate the fields.

The domestic and foreign political situation began to change drastically. In 1956 Indonesia unilaterally abrogated the Round Table Agreement. The following year Sukarno, outraged by the failure of the United Nations to act on his demands that it urge the Netherlands to negotiate with Indonesia on the return of Irian Jaya, seized all Dutch-owned business companies, banks, shipping lines, tin mines, and rubber plantations and expelled all Dutch citizens from the country. The loss of its professional managers was a severe blow to the fragile Indonesian economy, but Sukarno was launched on his doomed course to "rediscover" the revolution.

There was so much furor over the nationalization of Dutch properties that scant attention was paid in 1957 to the government's decision to assign jurisdiction over the embattled north Sumatran oil fields to the army chief of staff, General Nasution. It was a logical, political decision, since the area was in a state of anarchy and only military force and security could bring any order. Permina, the national oil company, was incorporated.

General Nasution tells the story that when he was considering whom to put in charge of the new company, he had three candidates on his staff who had in civilian life been a lawyer, an engineer, and a doctor. "I decided that an engineer would spend all his time planning what to do, and a lawyer would be too involved with what the legal problems were," General Nasution says. "However, a doctor's prime concern is to save the life of the patient, and that was what was needed. That is why I selected Colonel Dr. Ibnu Sutowo."

There were other reasons as well. Colonel Ibnu's unusual background, qualities, and experience had demonstrated that he was an expert organizer and logician. He was versatile, indefatigable, and impatient with any inefficiency. Above all, he was tenacious and an innovator. He had been an outstanding independence fighter and army commander in south Sumatra, in addition to his medical duties. The resolute young doctor tackled problems with such vigor and efficiency that his fellow officers called him The Tiger. Ibnu had wanted to be a doctor from the time he was in primary school. Since his father was a civil servant in Java for the colonial administration, he was one of the few Indonesians permitted to attend Dutch schools. He finished a nine-year university and medical course at the Netherlands Indies Medical School and on graduation in 1940 became a government doctor assigned to an antimalaria campaign in a colonization project of twelve thousand Javanese who had been resettled in Sumatra.

The malarial death rate in the area at that time was two hundred fifty per thousand, but in one year's time Colonel Ibnu reduced it to twenty-five per thousand. This was before DDT, and he achieved this astounding success by clearly diagnosing the problem and then doing something about it immediately, qualities that have characterized his entire career. His first action was to make food available to the people, because due to malnutrition they had no resistance to malaria's subsidiary diseases such as pneumonia and anemia. He made regular inspections of every house in the villages in his district, commanding the villagers to clean their houses and giving them an arsenic spray to try to break the mosquito-breeding cycle. He made the villagers take quinine, even holding their noses and putting the spoon in their mouths. The Dutch administrator told him, "Your program is too expensive for me." Ibnu replied crisply, "What value do you put on a human life?"

As a doctor, Ibnu had complete intellectual freedom and was totally independent in his attitude toward the colonial administration. The Dutch prohibited nationalist activities, but Ibnu, who was a born organizer, dispensed politics along with his quinine. He organized a youth political party and preached the need for independence. Being immensely practical, he also formed economic cooperatives to purchase and distribute commodities in the colonization project. During the Japanese occupation he became the civil leader of the project. Having hidden his medical supplies in small caches before the Japanese arrived in north Sumatra he was able to take care of his community until the end of the war. The Japanese found him as independent as had the Dutch. On one occasion he struck a Japanese officer who threatened him.

When the revolution began, Ibnu, at the request of independence leaders, took charge of Palembang, south Sumatra's largest city, and ran its two big hospitals. He joined the independence army as chief of health services for south Sumatra and was soon also a guerrilla fighter and army chief of staff for the area.

At the time General Nasution appointed him president-director of Permina, Colonel Ibnu was deputy army chief of staff in Jakarta. The imperturbable Ibnu, with no knowledge of the oil business and no funds at his disposal, had a practical philosophical approach to his new assignment. "There was a general belief, on the part of both the oil companies and the Indonesians, that Indonesians could not possibly manage an oil business," he relates. "This skill was somehow thought to be limited to foreigners. Therefore, at the beginning, the purpose of Permina as I saw it was not primarily to increase income for the state, much as that was needed, but to develop the necessary knowledge and skills of the oil industry."

Colonel Ibnu's oil company staff consisted of three army officers with administrative and purchasing experience. Initially they did not even have an office in Jakarta, and when they started to work in the oil fields their first headquarters was a wooden shelter built by a farmer to protect his cattle from monsoon rains. Their physical inventory was not much, either. They had a few old wells still seeping oil, rusted pumps, one leaky pipeline, and the heap of scrap iron which had once been the nation's first refinery. However, they had one piece of paper that was the key to being in business: a marketing contract to sell their oil.

The doctor who knew nothing about oil had the good fortune to meet the one type of oilman who could help him—an independent. Traditionally, independent oilmen have had to risk more and be keener entrepreneurs than the major oil companies with whom they compete. Harold Hutton, an independent California oilman whose company was Refining Associates Canada, Ltd. (known as Refican), had come to Indonesia looking for oil opportunities. He immediately recognized the potentialities of the infant national oil company and was ready to gamble on its success. "All it takes to build an oil company is oil," he told Colonel Ibnu. The two formed a bond of friendship which was a vital element of the enterprise.

A sticky question was whether Royal Dutch-Shell would legally claim any oil exported from its former properties. An answer was found during a catastrophe that struck the nation. In early 1958 the country was plunged into civil war. Dissident army and political groups in both Sumatra and Sulawesi, rebelling against Sukarno's ineptitude in handling the nation's affairs and the growing strength of the Communist party, proclaimed a new Revolutionary Republic of Indonesia. The unity of the Republic of Indonesia was at stake, and Sukarno mobilized seventy loyal army battalions that suppressed the insurgents after five months of all-out warfare. It was a bitter, ironic "war of brothers," as it was called, since both the loyal and the insurgent army forces were Indonesian.

The Royal Dutch-Shell properties were in the heart of insurgent territory, so to clarify legal title to the properties for Permina, General Nasution cancelled the Royal Dutch-Shell concessions under the State of War and Seige Law that was in effect. In any event the concessions would have expired in a few years. Royal Dutch-Shell had escaped being confiscated in the 1957 nationalization of Dutch properties because of its international ownership, and Dutch nationals were replaced with British and Americans in its management. The company made no efforts to protest the cancelling of the concession nor did it make any claims on the oil, as it was concerned about keeping its more recent concessions and hoped to be in a position to apply for new areas when a mining law was passed.

Despite revolutionary and guerrilla fighting, Permina, with Hutton's help, managed to rehabilitate some of the wells and repair the pipeline and storage tanks at the port. It was a "poor boy" operation of patching, mending, and scrounging through the ravaged countryside for scrap metal and parts, but on May 24, 1958, Permina triumphantly began loading a small coastal tanker with 12,000 barrels of oil to be sold in Japan. Indonesia's first commercial oil well was still producing after seventy-three years, and now it was appropriate that oil from this well was in the historic shipment celebrating the first business of what would soon become a billion-dollar company.

With the thirty thousand dollars realized on its first sale, Permina bought pipe, steel plates, pumps, and simple equipment. It was still a salvage operation, however. Army troops assigned to protect operations made a systematic search for scrap and found a treasure in a junkyard in Medan, the largest city in the area. The discovery was almost as important to Permina as its first oil shipment. The oil properties had been cannibalized over the years of war and revolution, and the junkyard contained such quantities of machinery, pipe, and oil-field equipment of various kinds, a great part of which was in good condition, that they filled fifteen boxcars.

Colonel Ibnu was beginning to build a staff for his oil company that would learn by doing. They patched together a teakettle refinery that could produce gasoline and kerosene for the local market and enable Permina to meet its payroll for the oil-field workers.

Ibnu was impatient with Band-Aid operations—he was after a major cure. He wanted ample capital to develop the properties with modern equipment and wanted Indonesians skilled in oil industry technology to do it. The opportunity to achieve these goals was not long in coming, and he was ready for it.

Shortly after the first sale of oil two groups of Japanese businessmen came to Indonesia to explore possibilities of getting a foothold in the oil business. Like Harold Hutton they recognized the potential of Permina and, after forming a consortium, began negotiations with Colonel Ibnu. In early 1960 an agreement was signed whereby the Japanese group extended a \$53-million credit in the form of equipment, machinery, materials, and technical assistance to Permina over a ten-year period, the loan to be repaid in crude oil out of increased production. Ibnu and the Japanese managed to overcome strong opposition to the agreement in both countries. Indonesian political and public opinion was antagonistic to all foreign investment. In Japan, where the approval of the government was necessary, there was serious concern over Sukarno's left-wing policies. Ibnu went to Tokyo, and the prime minister was so impressed with him as a young leader worth supporting that the government sanctioned the new agreement.

Although this venture was essentially a barter deal, it also contained a clause that was a first step toward what Colonel Ibnu would later use to develop a great international oil innovation–the idea of sharing production. As an incentive for the Japanese to help Permina increase production as quickly and substantially as possible he agreed to let them share any excess production over the amount of increased production agreed upon for the repayment of the loan.

Ibnu was also laying the basis for the development of Indonesian technological skills. In 1962 Colonel Ibnu established a Technical Training School in Pangkalan Brandan, the Sumatran oil center, to give oil-field workers refresher courses in oil-field and refinery technology and related courses. At the same time he founded a more advanced school at Bandung Technical Institute and began sending other young graduates abroad.

Colonel Ibnu drove himself as hard as he drove everyone else. His guiding principle has always been, "Learn while you work, work while you learn." He undertook intensive technical studies and in 1961 went to Saudi Arabia, Venezuela, Iraq, and Iran to learn the international politics of oil from their government oil leaders. He was impressed with the views of Venezuela's minister of petroleum, Juan Pablo Perez Alfonso, and Saudi Arabia's petroleum minister, Sheikh Abdullah Tariki. The year before, these two brilliant, fiery nationalists had founded the Organization of Petroleum Exporting Countries, known as OPEC, whose other members were the governments of Iran, Iraq, and Kuwait. At that time the five countries were supplying some 80 percent of all oil moving in international trade. Oil from these countries was produced and exported almost entirely by the big, private-enterprise, international oil companies who held concessions under a 50-50 profit-sharing agreement. Payments to the host governments, including royalties, were based on an income tax of 50 percent of profits reckoned at posted prices on all the oil sold. The governments of OPEC countries strongly believed that the oil companies artificially depressed world oil prices and that they were being swindled out of a fair return on the exploitation of their resources. Consequently the purpose of OPEC was to coordinate their oil policies and pressure the companies to increase oil prices and government oil revenues. Only Iran and Venezuela had formed government oil companies when Colonel Ibnu made his first trip, but another OPEC aim was to encourage the formation of national oil companies in all member countries in order to influence world oil markets.

Not even the architects of OPEC realized that after a decade of turbulent conflict they would surpass their most extravagant dreams and the balance of world oil power would shift from the international companies to the governments. However, at this point, in 1961, Ibnu's study of the complicated situation in OPEC countries had convinced him that Indonesia should join the organization to further its own goals.

Indonesia's long-awaited new oil and gas law of 1960 had been passed after nine years of study and indecision. In effect it nationalized the oil industry, since it stated that "the mining of oil and gas shall only be undertaken by the state" and "mining undertakings of mineral oil and gas are exclusively carried out by state enterprises."

How to reconcile this law with the existing concessions of the three big companies producing almost all of Indonesia's oil was yet to be determined by the government. The stage was set for a bitter three-year battle.

The law was an example of "socialism a la Indonesia," as Sukarno called his policies. It was not as arbitrary as it sounded, inasmuch as the law further stated, "The minister may appoint other parties as contractors for the state enterprises, if required for the execution of operations which cannot or cannot yet be executed by the state enterprises involved as holders of the authorities to mine themselves." In an explanatory memorandum issued with the law, the government frankly recognized that "the mineral oil and gas industry requires a very large capital outlay as well as a great and extensive skill regarding the branches of mineral oil and gas production, so that according to this regulation it is still possible for foreign enterprises to work in Indonesia as contractor of a state enterprise, on conditions which will satisfy them."

The crucial question was, what "conditions" could Caltex, Stanvac, and Royal Dutch-Shell negotiate with the government. Whatever they agreed to in Indonesia was bound to affect their worldwide holdings where there was far more at stake in terms of known oil reserves. Under the new law the companies were no longer concession owners of national oil in the ground; but what was a "contractor"? The protracted dispute that developed centered on the companies' insistence on retaining management control of operations and the exclusive right to set oil prices and on their reluctance to meet the government's demand of 60 percent of net profits.

After almost a year of deadlock, the government unilaterally settled two of the issues by a presidential decree, setting its profit share at 60 percent and stating that the government would participate in determination of selling prices of crude oil and refined products. Negotiations over management control continued to drag on.

Although the political climate in Indonesia in 1962 was not favorable for new foreign investments, the lure of Indonesia's great oil potential, which had been locked up for so long, attracted other big oil companies. Pan American Oil Company, a subsidiary of Standard Oil of Indiana, was anxious to enter the international oil industry and had no big foreign holdings at jeopardy. It decided to take the new law at face value and become a "contractor." In June 1962, after two whirlwind months of conferences and while negotiations with the three big companies were still in progress, Pan American signed a Contract of Work on the formula of 60-40 profit sharing on realized market prices. It would be a contractor on a thirty-year agreement with a $28.25 million exploration commitment over eight years on 9 million acres in central Sumatra. It also paid a $5 million bonus, which the government badly needed, and agreed to pay additional cash bonuses as various production levels were reached. The state enterprise was tax collector and auditor and had the right to elect to take up to 20 percent of oil production in kind to sell for itself.

It was a black day for Caltex, Stanvac, and Royal Dutch-Shell. Precedents had been set. However, their cases were much more complicated, since they involved agreements on established production, refining, and domestic marketing in addition to new exploration areas. Negotiations had begun in late 1960. By April 1963, with no agreements in sight, tensions had mounted to the breaking point. President Sukarno issued a regulation that if new agreements were not reached on a specific date, seven weeks away, the government would take unilateral action.

The ultimatum was a bombshell in capitals around the world. If American, British, and Dutch interests were ousted, it would mean that Indonesia would turn to the Soviets for petroleum development and technology. It would mean a major break with the Western world. President Sukarno was also heating up his "Crush Malaysia" campaign. President Kennedy offered to help mediate the dispute, as far as the American companies were concerned, through a team of U.S. government advisors. President Sukarno agreed and suggested a meeting in Tokyo. Wilson Wyatt, lieutenant governor of Kentucky, represented President Kennedy, and the rest of the team included Howard P. Jones, U.S. ambassador to Indonesia, Walter J. Levy, an internationally known oil consultant, and Abram L. Chayes, legal advisor to the Department of State. President Sukarno was accompanied by Chairul Saleh, third deputy prime minister and minister of basic industries and mining, and Lukmam Hanafiah, a government legal adviser on oil. The parent companies of Caltex and Stanvac sent attorneys. Royal Dutch-Shell, which obviously would have to go along with whatever was decided, sent observers to this American-Indonesian conference.

The three-year controversy was settled in three days, and it seemed that a new era had dawned for Indonesian oil. All three companies gave up their concessions and became contractors on essentially the same terms as Pan American. They were given twenty-year exploration contracts for their former concession areas and thirty-year contracts for new areas. Each company selected areas near its existing productive areas, made exploration commitments totaling approximately $30 million, and paid a $5-million case bonus. The financial risk of exploration was viewed as part of the contractor's normal business risk in consideration for receiving an exclusive contract. A gradual take-over of existing marketing and refining facilities, with compensation to be paid to the oil companies, was agreed upon. Marketing was to be nationalized in five years and refining in ten years, as the Indonesians felt this would give them time to train their own management.

In addition to Permina, there were at that time two other state oil enterprises, Pertamin and Permigan. Pertamin had been formed to handle production from a joint government-Royal

Dutch-Shell concession in which Royal Dutch-Shell's interest reverted to the nation. Its Indonesian staff had been trained by Royal Dutch-Shell to take over. Permigan operated two small fields and an old refinery in central Java which the government bought from Royal Dutch-Shell and which the company was glad to get rid of since the anti-Dutch and Communist-dominated workers had always been a source of trouble and had threatened to seize the properties. Since the foreign oil companies were required to make their Contracts of Work with a state enterprise, Caltex was assigned to Pertamin, Shell to Permigan, and Stanvac to Permina. Pan American had already been assigned to Pertamin. It was a cumbersome, inefficient method of handling matters. Although there was a general management board, there was little coordination between the state companies.

It was a relief to everyone to have the matter settled and to establish an atmosphere of cooperation rather than conflict. "Now we can roll up our sleeves and get to work," a Stanvac executive remarked. The prospect of new exploration was exciting. For the first time, exploration techniques and thinking developed in the past quarter-century could be applied to new areas. For example, as yet no deep wells had been drilled in Indonesia. The government was elated at the prospect of increased revenues.

However, the dark shadows of coming events were already cast. An ominous incident occurred on the day oil company and government officials were holding a final conference on details of the work contracts. They were assembled in a conference room of the luxurious new Hotel Indonesia. In midafternoon the conference recessed briefly, and the participants silently watched from the windows while a shouting, tumultuous, Communist-organized mob set fire to the British Embassy across the traffic circle. Finally, one of the oil company officials suggested they get back to the business at hand. The conference proceeded as though nothing had happened.

Sukarno's "Crush Malaysia" campaign was to be the death knell of Royal Dutch-Shell in Indonesia. The island of Borneo was the potential battlefield, since North Borneo was part of Malaysia whereas the rest of the island was the Indonesian province of Kalimantan. Almost half of Royal Dutch-Shell's production of 100,500 barrels a day came from this area, and in 1962 it had finished a crude-oil pipeline to carry 40,000 barrels a day through the jungle to its terminal and refinery at Balikpapan. The region's military commander was sympathetic to the Communist party and permitted the Communist-dominated labor union to harass foreign personnel and upset operations to the point where it was virtually impossible for the company to continue operations. Royal Dutch-Shell threw in the towel in mid-1965 and began negotiations with the government to sell all its Indonesian assets.

The Communist party clamored for the nationalization of all foreign companies and did everything possible to disrupt their operations through wildcat strikes, sabotage, and inciting public opinion against them. Stanvac decided to lessen its frustrating burden by offering to sell the government its Sungei Gerong refinery in south Sumatra and its assets in several adjacent fields in which production was declining.

Only Caltex, which was producing 50 percent of Indonesia's oil, managed to survive relatively unscathed. Its location in the remote jungles of central Sumatra, away from any center of political activity, and the fact that it did not have any refining or marketing activities were greatly responsible for its ability to continue operations.

In early 1965 President Sukarno appointed Colonel Ibnu a minister of state to advise Deputy Prime Minister Chairul Saleh in government oil and gas affairs. Ibnu worked out the details of purchase of the Royal Dutch-Shell and Stanvac properties. The government had realized since the 1963 agreement with the oil companies that the three state oil companies should be unified to deal with the contractors. Now, with the unexpected purchase of properties, the need for this was even more obvious.

Under Colonel Ibnu's direction, Permina was expanding rapidly, producing and exporting 21,000 barrels a day. He had made contracts with Harold Hutton's Refican to rehabilitate old wells and explore new areas in north Sumatra. He also signed an exploration contract with Asamera Oil Corporation, Ltd., another small Canadian independent. With them he established the 60-40 profit split, even before he did so with Pan American. Ibnu's company had only 5 percent of Indonesia's oil production, but Ibnu, who had become a general, was now the most experienced and aggressive Indonesian oilman. However, he was uncertain about his future: he was an avowed and outspoken anti-Communist yet his country had gone the other way. But it was the eve of September 30, 1965.

5
"Share Oil, Not Money"

On the morning of October 1, 1965, when the Communists launched their abortive coup, General Ibnu was in complete charge of the nation's oil affairs. Deputy Prime Minister Chairul Saleh, who was abroad, had temporarily appointed him minister of oil and gas. While General Suharto was establishing order, General Ibnu's first act was to ban Perbum, the Communist oil-field-workers' union, from any activities in government oil fields. Although it would be almost five and a half months before circumstances would give General Suharto the power to outlaw the Communist party, General Ibnu moved swiftly to eradicate their control of Permigan, the government's small oil company in central Java, an area that was a Communist stronghold. Later, General Ibnu learned that his name was high on the list of generals to be liquidated if the coup had been successful.

General Ibnu had a unique and appropriate solution for the troublesome Permigan company. As a commercial venture, it was an uneconomic operation of old fields and a primitive refinery. However, the facilities offered an exceptional opportunity for the establishment of an academy where students could be trained according to his favorite precept: learn while working and work while learning. He dissolved the company and, in early 1966, gave the fields and refinery at Cepu to the national research Oil and Gas Institute to establish an Academy of Oil and Gas. He also planned to use the new academy as the stage for a drama that he knew was in the making.

Despite the chaos into which the country was thrown, General Ibnu had concluded negotiations with Royal Dutch-Shell for the purchase of all its assets. It was a major acquisition, as Royal Dutch-Shell's production of 100,500 barrels a day from its contract areas in south Sumatra, Kalimantan, and Java was almost a fourth of Indonesia's total and the company had three refineries. It had 11,000 employees, three-fourths of whom were Indonesians. General Ibnu convinced Royal Dutch-Shell's negotiators that the price should be one that Indonesia could realistically hope to pay out of its profits from operation of the properties. When the negotiators suggested $150 million, he offered $110 million, which they accepted, and an agreement was signed on December 31, 1965, transferring the properties to Permina.

There were many, both in Indonesia and in international oil circles, who predicted that Indonesians would be unable to operate such a major enterprise successfully and that it would not be long before operations would break down. They underestimated the basis for success which General Ibnu had already established in building Permina from nothing. He and his staff had learned the oil business from the bottom up and were convinced they had the knowledge and ability to run the company smoothly. Furthermore, he had the resources now to begin to put into effect the ideas and plans he had been formulating concerning the true role of a national oil company.

General Ibnu knew that one of the most important aspects of good management is the morale and beliefs of its employees. During the years of bitter conflict between the foreign oil companies and the government, morale had deteriorated greatly. However, his greatest concern was the extent to which the thinking of all workers, from the oil fields to the offices, had been affected by Communist-party activities and propaganda and by Sukarno's "socialism a la Indonesia."

"Ideology is an important, decisive factor in the endurance of the state," General Ibnu told me shortly after he took over.

> The influence of Communist thinking goes deep. For fifteen years Indonesia had been indoctrinated by the Communists. Stopping the Communist efforts in 1965 did not mean that all that had been planted for such a long time, in a very systematic way, would disappear. After fifteen years of communism, people can't help thinking like Communists when it comes to economics. My job was to do something positive to help people change their thinking and learn how to think along democratic lines. I wanted to show Permina's employees why communism doesn't work. I also wanted to know how many Communists I had in the organization, as many of the engineers were trained in Moscow.

To achieve this he initiated a bold, unusual training program. At the new Cepu Academy of Oil and Gas he set up a three-week "mental upgrading" course that all of Permina's employees, now increased in number from four thousand to fourteen thousand, were required to take. Cepu was an isolated spot with no distractions where the employees, in small groups, could study intensively, away from their families and jobs. The course taught them specifically why Marxist thinking was wrong and what it had done economically to Indonesia. They were told exactly what the Communist party's plans had been. They were given an understanding of how a citizen of a democratic country should think and feel and what Indonesia could hope to accomplish now. Also, they were shown the importance of their individual jobs in rebuilding the nation. To the extent the oil industry improved rapidly, so would the country as a whole.

General Ibnu fired the first employee who refused to take the course. "After that, I had no more trouble," he commented. "By the time half of the employees had been through the course there was a marked difference in their attitude and way of thinking. It improved their efficiency and their dedication to their jobs."

The course was so successful that other government departments set up similar courses. Ibnu, the doctor, had diagnosed a chief cause of national cancer and shown how to treat it.

In March 1966, when Sukarno transferred his powers to General Suharto, General Ibnu was appointed minister of mines and minister of oil and gas, in the new Suharto cabinet and initiated the trend of events which would change Indonesian economic history and the course of international oil development.

It was obvious that one of the most important ways to generate new income for the country was to encourage foreign investment in the search for new oil resources. Geologically, Indonesia's prospects were luring. Since the end of World War II, oil technology for the exploration for oil under shallow coastal waters had advanced greatly. Geologists believed that Indonesia's 3 million square miles of shallow ocean waters could cover many big oil fields. However, under the oil nationalization law and in the new political situation, any newcomer to the Indonesian oil scene would have to work as a contractor to Permina. The welcome mat was out, but from the viewpoint of big international oil companies it was a prickly one. Now that General Ibnu was in charge there was no question that the conditions he would demand of contractors would be different than those which had been negotiated with the three big oil companies following the 1963 Tokyo agreement for their Contracts of Work.

General Ibnu had been outspoken in his criticism of these contracts at the time they were negotiated. He called them "dressed-up concession agreements."

> The most important thing in the difference between a concession and a nonconcession is the matter of ownership

of the oil. The fundamental principle underlying a concession, and highly favored by foreign oil companies, is that oil is owned by the government in its natural geological form, but as soon as man has done something to it, he is the owner of the oil. In other words, oil at the well-head becomes foreign property and the company pays a tax in the form of 60 percent of the profits from the sale of the oil. Under the work contracts, whenever we want oil we have to request it from the foreign oil company, and we can ask for and obtain a maximum of only 20 percent of gross production. These two aspects indicate that the work contract is no better than the old concession, and this should be opposed. Furthermore, if we are to adhere to the principle that the oil belongs to the Indonesian people, then, first of all, the management of the oil fields must be placed into Indonesian hands.

Ever since his studies of international oil politics and policies in the Middle East, General Ibnu had been convinced that the way to eliminate the conflicts of interest between foreign oil companies and host governments was to establish a production-sharing system to "share oil, not money," as he put it, and for the national oil company to retain management. Sharing of oil would eliminate the constant friction over whether the price of oil used to figure the profit split was artificial or was manipulated by the companies to the government's disadvantage.

He had used production-sharing concepts in the contracts Permina made in 1961 with the two small independent companies, Refican and Asamera, which had not yet been legalized by congressional ratification. Now the big question was, would this principle prevent the generation of oil investment capital on the huge scale the nation required to develop its potential as rapidly as possible. Major oil companies were adamant in their opposition to relinquishing management control, regardless of how tempting Indonesia's offshore potential was.

Fortuitiously, now as at Permina's inception independent oilmen were on hand to help pioneer an answer. A small, venturesome group of successful independent Colorado and Montana oilmen had formed Independent Indonesian American Petroleum Company, known as IIAPCO. They had first come to Indonesia in 1964 hoping to get a toehold as a contractor under the new law; they had selected an offshore site but had been unable to reach a satisfactory agreement with the Sukarno regime. As soon as the group learned that General Ibnu was receptive to proposals for offshore contracts on a production-sharing basis, their representative, Donald Todd, arrived in Jakarta with full authority to negotiate. The success of independent oilmen in foreign operations is based primarily on their ability to recognize opportunities in a new oil area or situation, their willingness to accept stiffer initial terms than will major companies and, most of all, their flexibility in being able to make quick decisions. They are not hampered by the time-consuming procedure of clearing policy decisions through committees and boards of directors which major company negotiators must follow. Donald Todd did not hesitate to accept all of General Ibnu's basic conditions to establish the international oil industry's first production-sharing contract.

In explaining his concept of Indonesian management control, General Ibnu told Todd: "This doesn't mean that we insist on making every decision, but we do insist on being able to make any decision we find necessary." Permina would approve IIAPCO's work proposals in advance, and the spirit of the contract would be one of partnership. It took less than two months to draft the terms. It was a thirty-year contract with a profit split, in terms of crude production, of 65 percent to Permina and 35 percent to IIAPCO. The contract area covered 21,150 square miles off the shore of northwest Java, and IIAPCO guaranteed to spend $7.5 million in exploration, at its own risk, within a period of six years. If oil was discovered it would recover its exploration cost out of a maximum 40 percent of production before profit sharing. All equipment that IIAPCO brought into Indonesia would become the property of Permina, but its cost also could be recovered from production.

Although the contract drafting went smoothly and quickly, there was stormy weather ahead. At the end of July 1966, four months after his appointment as minister of mines and of oil and gas, General Ibnu resigned. General Suharto was forming a new cabinet following the convening of the congress that confirmed the transfer of Sukarno's powers to him. General Ibnu had to choose between remaining minister in the new cabinet or giving up his position of president-director of Permina. He felt that he could do more for Indonesia by concentrating on developing a strong national oil company. He did, however, retain a key policy position as director-general of the ministry's Oil and Gas Affairs Department in the new Ministry of Mines.

The new minister of mines, Ir. Bratanata, knew little about oil. He was a mechanical engineer who had held various high positions in the government and in state enterprises for many years. He did not agree with General Ibnu that it was essential to establish the principle of production sharing for Indonesia's best long-range interests. He considered that profit sharing of money from the sale of oil, as established in the Contracts of Work, was adequate and that the quickest way to obtain oil development was to divide new exploration areas into small blocks and put them up for competitive bidding with cash bonuses on the signing of contracts.

For five months a period of Indonesian dualism existed in regard to the nation's oil policy. General Ibnu signed the IIAPCO contract as president-director of Permina and director-general of the Oil and Gas Affairs Department, at the same time advising Todd that it would need General Suharto's approval. Within the next three months he signed three more production-sharing contracts, on the same terms as those with IIAPCO–one with Refican for an offshore area southeast of Kalimantan, one with Japan Petroleum Exploration Company (with Japanese government participation) for an offshore area east of Kalimantan, and one with Kyushu, another Japanese company, for an offshore area south of Kalimantan. In the meantime, Minister Bratanata was recommending to General Suharto and the cabinet that Contracts of Work be made and was drafting regulations for them. A large number of major oil companies were negotiating with Minister Bratanata and backing his ideas in the hope that the principle of production sharing could be defeated and that the choice areas granted to the four small companies would be thrown open for bidding.

The climax of the contest came in January 1967, when Minister Bratanata, who had submitted to the cabinet his final regulations on the division of offshore areas, sent a letter to interested oil companies offering seven offshore areas, including those on which General Ibnu had made production-sharing contracts. General Ibnu knew how to present his case. Two days later Minister Bratanata was advised by General Suharto that he was approving the production-sharing contracts. Policy decisions for the Oil and Gas Affairs Department were put directly under the cabinet.

From the beginning, General Ibnu had predicted that if he established the principle of production sharing, the big companies would eventually accept it. There was no great rush at first. A number of large companies initiated protracted negotiations. The management control clause continued to be an obstacle that every company tried to have deleted or modified. General Ibnu would not compromise. Finally, in late 1967, Continental Oil Company became the first major oil company to sign a production-sharing contract. Shortly after, two more majors, Union Oil Company and Sinclair Oil Company as partners, fell into line. Negotiators for these companies explained to me that their decision was based on the fact that as long as their contracts permitted them to make exploration expenditure decisions they were not unduly concerned about future implications. During the course of their negotiations they had developed such respect for General Ibnu that they felt that if production were developed they would be able to work out satisfactory operating agreements. This faith was more than justified, as the future would prove.

With the logjam broken, Hotel Indonesia was filled with major oil company representatives and independents, all competing for contracts. The poker stakes went up. General Ibnu made it clear to the newcomers that each contract would have to be more favorable, in some way, to the government than the one signed before. He was a shrewd bargainer. Continental paid the first signature bonus, of $1 million, and when it signed another contract the following year it paid a bonus of $7 million. During 1968 General Ibnu signed fourteen production-sharing contracts with such major companies as Mobil, Gulf, Phillips, Superior, Compagnie Française des Pétroles, and Agip, of Italy. In this one year alone he obtained more than $18.3 million in signature bonuses and $54 million in production bonuses to be paid when production reached levels ranging from 50,000 to 300,000 barrels a day. In the majority of the 1968 contracts there was a new clause stipulating that when production reached 65,000 or 70,000 barrels the government's production share would increase to 67½ percent. However, the most important proof of the success of his production-sharing principle was that in Permina's first twenty contracts he had generated a total commitment of $220.5 million to explore Indonesia's potential, most of it offshore. Pan American had relinquished its Sumatran contract in 1967 after spending almost $28.3 million on fourteen dry holes, but the oil industry is a gambling game noted more for its losses than its successes. The point is to be able to stay in the game, and Indonesia now had a seat at the big table with a stack of blue chips that she didn't have to pay for, and ownership of the oil would be hers if it was there.

In mid-1968 General Ibnu achieved another of his most cherished goals: the unification of state oil enterprises into one national oil company in charge of all the nation's oil affairs. Suharto, now acting president, had appointed a new minister of mines—Sumantri Brodjonegoro, the president of the University of Indonesia and a chemical engineer—who was greatly in favor of unification. Policy matters in connection with the Oil and Gas Affairs Department were put back under the Mines Ministry, and Permina and Pertamin, which primarily handled marketing, were merged into Pertamina. General Ibnu was president-director of the new company.

In East Kalimantan, seismograph parties cross rice paddies such as this. Pertamina's discovery well, in August 1972, of the onshore Sangatta field 125 miles north of Balikpapan, flowed 2,200 b/pd from zones between 3952 and 4048 feet.

Geophysical crews in North and Central Sumatra find it heavy going through the jungles, but are finding promising new structures.

It was just two and a half years since General Ibnu had bought the Royal Dutch-Shell properties, and among those who had predicted that the national oil company would fail were some who now were the most eager to go to work for it as a contractor.

Just as President Suharto had assembled a team of the country's most brilliant economists to work on overall economic problems, so General Ibnu assembled a team of outstandingly competent and experienced Indonesians to manage and expand Pertamina.

They were ready to capitalize on all the foundations so painstakingly laid. One of General Ibnu's early major breakthroughs had been the founding of the Far East Oil Trading Company in 1965, a joint venture of Permina and twenty-one Japanese shareholders representing independent oil companies, utilities, and industrial users. He had recognized the importance of creating a new market in an oil-dependent Japan in order to compete with the major companies, once he had the oil to do it. Far East sold only 3.5 million barrels in 1966, but when it hit on the idea of persuading the Japanese government and electric companies to mix Indonesian low-sulfur crude with Middle Eastern high-sulfur crude to reduce pollution, sales began to climb. Far East sales reached 13 million barrels in 1968 and would quadruple that figure in the next two years.

In 1967, the year in which General Ibnu was trying to promote more interest in production-sharing contracts, he established an office for the national oil company in New York City to make information about the Indonesian oil situation available to interested companies. "It will save them such a long trip coming out here," he said, "particularly when they find after they get here that we insist on something they wouldn't be interested in–management control." However, he had long-range plans. He wanted the company to be close to the center of international oil and financial operations.

In Indonesia, national oil company production had risen from 36,618,000 in 1967 to 37,111,000 barrels in 1968. Pertamina had discovered two important new oil fields, one in southern Sumatra and one in Kalimantan. Its own exploration program was beginning. Caltex, which was producing 70 percent of Indonesia's crude, increased its production by 20 percent to 164 million barrels that year. It looked forward to further big increases, as it was accelerating its exploration program at the rate of $6 million a year.

For Indonesia the momentous oil events of 1968 were already beginning to prove that the hopes of President Suharto and his economic team would be fulfilled. However, most exciting and portentous of all was that as the year ended drillers of Indonesia's first offshore wildcat well, on IIAPCO's production-sharing area, were already grinding a hole deep into the rocks beneath the waters of the Java Sea. A multimillion dollar question was about to be answered.

6
The New Order Comes of Age

On August 16, 1970, President Suharto put the nation on the psychiatrist's couch. The occasion was the annual Address of State to the House of People's Representatives on the eve of Independence Day. The address was of special importance, as it marked the celebration of Indonesia's twenty-fifth anniversary as an independent nation. It was also the fifth year of the New Order and the end of the first year of the Five Year Development Plan.

As always, the president spoke frankly to the people. "Only part of us—just the smaller part—have enjoyed the fruits of independence," he said. "The greater part, however—soldiers, civil servants, small peasants, and laborers—still live a difficult life. They still live in inadequate houses in alleys and dirty slums without fresh drinking water and without electricity, with a low degree of nutrition, deeply concerned about the schooling and education of their children as if looking forward into a dark future. Why, after being independent for a quarter of a century, are we still living under such conditions?"

The answer to this question, he told them, was that the nation had never implemented economic development "in the true sense of the word." Step by step, he reviewed the errors of government from 1945 to 1960, when the nation entered a five-year period that he characterized as a "life-and-death struggle, a struggle for survival and to uphold Pancasila and the 1945 Constitution."

It is impossible to understand Indonesia today without an appreciation of the role of Pancasila in molding the nation's spirit and outlook on life and of its profound significance as a philosophical guide to government actions and policies. *Pancasila* is Sanskrit for Five Principles, and it represents the five inseparable and mutually qualifying principles of the nation's official ideology. These were first proclaimed by Sukarno in a speech on June 1, 1945, to the committee of nationalists who were preparing for the proclamation of independence which came two months later. Sukarno did not claim to have created the Pancasila. He said he "dug deep into the soil of Indonesia's traditions" and communed with God for "incalculable hours" while he was imprisoned, and formed what he claimed was a "God-sent inspiration" for the principles upon which to base the republic.

The principles are a mixture of democracy, nationalism, internationalism, Islam, and mysticism, which have deep meaning for all Indonesians. They are: *Belief in the One Supreme God.* This foremost principle, embodied in the constitution, is the basic foundation of all religious philosophies found in Indonesia. Every person is free to worship according to his own conscience. (2) *Belief in a just and civilized humanity.* This affirms respect for all peoples of the world and the desire to unite with the peoples of all nations to promote the cause of world peace and oppose actively all forms of imperialism and colonialism wherever they occur. (3) *Belief in nationalism.* This proclaims the national unity of Indonesia and the principle that nationalism is a liberating movement as opposed to imperialism and colonialism. (4) *Belief in democracy.* This is based on the belief that democracy is the natural condition of man, although it is modified to fit particular social conditions. Indonesian democracy is based on two ancient Indonesian philosophical concepts: *gotong-royong,* or mutual assistance; and *musyawarah,* holding deliberations, to achieve *mufakat,* or unanimity. Democracy is also the principle of representative government with sovereignty vested in the people and exercised through their representatives in the People's Consultative Assembly. (5) *Belief in social justice.* With social justice Indonesians link social prosperity, because they regard the two as inseparable, believing that only a prosperous society can be a just society, although prosperity itself can reside in social injustice. Belief in social justice is also belief in the social, economic, and political equality through which people work together, by *gotong-royong* and *musyawarah*, to secure prosperity for all.

The people accepted the Pancasila as the true expression of their innermost beliefs of how the nation should conduct its political, economic, and social activities. However, as Sukarno veered toward communism he began to bend the interpretation of the Pancasila to suit his own political adventuring, and the struggle to uphold the Pancasila evolved. When the leading army generals were murdered during the abortive Communist coup, they became known as Pancasila Heroes of the Revolution. President Suharto proclaimed October 1st as the Day of the Inviolability of Pancasila, to be commemorated annually for the success of the Pancasila forces in suppressing the September 30th Movement of the Communist party, because its aim was clearly to abolish Pancasila. National observance of this day is intended to heighten vigilance in safeguarding the Pancasila state philosophy and to fortify the nation's resolve to practice it in all ways.

In his twenty-fifth-anniversary speech President Suharto pointed out to the people that Sukarno's deviation from Pancasila had led them to a democracy that in practice was very near to the system of dictatorship. "Pancasila is not just an empty phrase," he said. "Pancasila is the complete outlook on life of our nation, the crystalization of the identity of our nation and at the same time the form of the way of life we desire. The New Order constitutes a total correction of deviations in all fields occurring in the preceding times. It is an effort to rebuild the power of our nation and determine proper ways to develop long-term national stability, so as to speed up the process of nation building based on pure implementation of the Pancasila and the 1945 Constitution."

In presenting the balance sheet of the New Order's achievements and failures, President Suharto pointed out that:

> This was not patchwork, but a complete overhaul. The condition of our economy is now better. All of you have experienced that food and clothing are in sufficient supply and at stable prices. We do not have to queue for rice, petroleum, sugar, or salt. Prices no longer jump. Communications within the city, intercity, and interisland are smoother by land, sea, and air. Not only in the cities but also in the villages there is activity and enthusiasm to work to improve and increase production. I sometimes have to make difficult and unpopular decisions, but all these things are the results of our joint efforts. It is by your sacrifices that we were able to achieve these results.

Although much of the first year's energy in implementing the Five Year Plan was spent in establishing the plan and setting up procedures, the wheels of development were already turning at an accelerated rate. The value of exports, including petroleum, reached more than $1 billion, almost 10 percent higher than the target fixed for the first year and an increase of a little more than 16 percent over the year before. Rice production exceeded the target, and output of a wide variety of other agricultural products increased. Textile production was 30 percent higher. Bank credits, foreign and domestic capital investments, foreign aid, and tax revenues all greatly increased.

One of the capital investments that President Suharto stressed as having great significance in demonstrating the people's enthusiasm for work was represented by what village communities had achieved. He set up 65,000 village projects and gave each village the modest sum of $300 to be used by the villagers to help themselves. The results were remarkable. A total investment of $12 million produced improvements valued at $34 million. The projects included 24,000 irrigation systems to provide water for 1.8 million acres of rice fields and construction of 4,000 new village rice barns, 16,500 new village bridges, 4,000 rehabilitated village markets, and 3,000 new village meeting halls, schools, religious schools, and mosques. This revitalization among the rural people of *gotong-royong,* mutual assistance, is firing the people's development spirit and carrying over into all areas of their activities.

Although the government's emphasis was on economic development, President Suharto reported progress in other areas as well. The quality of primary school education was improving with the distribution of 3.3 million books. Vocational and teacher-training schools had been rehabilitated, as had hospitals, clinics, and mosques.

President Suharto suggested that the best way to assess the progress of the New Order's first five years was to apply a "very simple yardstick" of how it had met the "people's demands." When it was initiated at the end of 1965 there had been three demands: dissolution of the Communist party, the purging of pro-Communist and incompetent government officials, and the lowering of prices. In 1967 the people, led by university and high school students, changed them to two demands: the lowering of rice prices and the eradication of corruption. Now, in mid-1970, Suharto pointed out, the people, led once again by students, were making only one demand: "that the eradication of corruption be stepped up."

"In this matter of eradicating corruption we are of one opinion," the president asserted. "The students, the press, the public, all righteous people, I myself, have one common determination: corruption has to be eradicated. It is detrimental to the state finances, which means it is detrimental to the people. It constitutes a threat to development. It is a violation of the law and is against one's morals and feelings of justice. I myself will lead the fight against corruption."

Indonesian students have great political influence and have been a major organized factor in politics since the early days of the nationalist movement. Student revolutionary groups were at the forefront in the fight for independence. Sukarno had believed Indonesian youth to be his strongest followers, but, following the massacre of the generals, moderate university and high school students, encouraged by the army, formed two strong national youth federations that played an aggressive, successful part in destroying the Communist party. They turned on Sukarno and played a major role, by their demonstrations and demands, in ending the governmental dualism of power between Sukarno and Suharto. The New Order is sensitive and responsive to the student groups, which, in effect, are self-appointed political watchdogs. They have the power to arouse and lead public opinion.

In January 1970 Jakarta retail prices rose 7 percent following an increase in the market prices of rice and the administered prices of petroleum products. This sparked serious student agitation and demonstrations against government corruption. The students claimed that the price increases of gasoline and kerosene were not justified in view of "huge profits" made by the state oil company, Pertamina, which they asserted had grown so big that it was beyond government control. They also charged that BULOG, the state rice distribution agency, and the state forestry enterprises were engaged in corrupt practices. Student demands to stamp out all corruption were backed by the press, prominent intellectuals, and a number of respected political figures.

President Suharto responded to student pressure by establishing a Committee of Four to investigate corruption in all its forms and to present recommendations to the People's Consultative Congress for new laws to eradicate corruption and control the activities of state enterprises. Furthermore, he unexpectedly invited representatives of some twenty student groups to the presidential palace for a three-and-a-half-hour exchange of views. This ended with his inviting students and others with solid evidence about corruption to visit him each Saturday morning at his home. In discussing the benefits of this approach President Suharto told me, "It also gives me the opportunity to correct them if they have false information and to give them a full understanding of the dedication that government leaders have to building this nation."

President Suharto's move to make an end to corruption, and his own personal responsibility, took an unusual turn when he learned that some government and army officials were taking second wives. Indonesia is predominantly a Muslim country, and religious law permits more than one wife. However, to President Suharto, as an economist, it was a sign of corruption, since an official obviously could not support more than one family on a government salary. He had a thorough investigation made, and a number of army officers, and officials, were politely given the opportunity to resign.

The student groups, convinced that President Suharto was taking proper measures, decided to dissolve their Anti-Corruption Committee on the night before the president's twenty-fifth-anniversary report to the nation. However, they planned to assemble for a *tirakatan* night, or night of meditation, to demonstrate their feelings. Ali Sadikin, governor of Jakarta, to avoid a demonstration that might get out of hand, persuaded the students to cancel their plans and, instead, to stay indoors and join all citizens in a *tirakatan* to be observed by switching off their lights from 11:00 to 11:15 p.m. He suggested that during this blackout the people "pray to God by asking forgiveness for all wrongs and deviations and ask God's blessing on all national leaders who have resolved themselves to eradicate corruption."

Almost one-fourth of the president's anniversary report was devoted to the problems of eliminating corruption, the steps the government had taken, and the report of the Committee of

Four. "We have to be realistic and pragmatic in tackling this problem and make preventive approaches," he said. "The economic approach has our priority."

Much progress had been made in stamping out smuggling of exports and imports, black markets, and hoarding. The president pledged a drastic overhaul of state agencies to eliminate opportunities for corruption and in intensification of anticorruption investigations to apprehend and try offenders.

In analyzing Indonesia's problems and goals for the 1970s, President Suharto defined the long-term, basic objectives as continued concentration on economic development and development of political and economic democracy based on Pancasila. In the short-term period the two basic tasks, which had to be accomplished simultaneously, were to make the general elections in 1971 a success and to implement the Five Year Development Plan successfully. Suharto reminded the people that "the democracy we are cultivating is not a 'formal' democracy nor a democracy 'forced' upon us from above, but a democracy that grows and has its roots in society."

This was not idle oratory. Suharto demonstrated his concern about democracy at the rice roots by increasing his unannounced "silent inspection" trips. "I want to be my own eyes and ears in seeing the progress our development program is making," he told General Bardosono, one of his administrative aides. "I want to find out from the people what their problems are and what they are thinking. I don't want anyone to know I'm coming and I don't want anyone in the government to know where I've gone or why."

These instructions posed a formidable security problem. On one week's trip to cover dozens of villages in Java, General Bardosono visited the villages beforehand, simply telling the chiefs that a "VIP" was coming and giving them a plan for stationing police on the roads. Having been with Suharto for thirty years, he knew it was useless to plan details, as the president would create his own plan as he went along.

President Suharto dressed in farmer's clothes sowing rice.

The official party was small, all in civilian clothes, and traveled in an unmarked little bus. As the tour proceeded, a pattern began to evolve. In each village the president preserved his anonymity as long as possible, visiting farmers in their fields and going through crowded marketplaces examining vegetables, fruit, and meats and asking questions constantly about prices, conditions, and problems. When eventually he was recognized, he would invite the villagers to meet with him. Flustered village officials would hastily set up a public-address system and Suharto would speak at length, telling the people the things they needed to do in agriculture and that they would be shown films to teach them new techniques. He would tell them of the growing spirit of development throughout Indonesia and that the improvements they could see in their own village were happening everywhere; and that although life was still difficult and there were many problems yet to be solved, the whole mental attitude of the country had changed and they were no longer living from crisis to crisis.

President Suharto's deep understanding of the people and village problems stems in great part from an unusual opportunity he had when he was fifteen. Suharto's father, who worked for the government in water distribution for the rice fields, sent him to live with friends in a larger village so that he could go to school. There he met Kiai Darjatmo, a famous Islamic religious teacher and mystic, who became so interested in the young Suharto that he invited him to live with him and be his assistant. Darjatmo was also a water distribution official and a *dukun*—a medicine man and seer. Darjatmo traveled to many villages, and the peasants sought his help for all their problems, spiritual, physical, family, and farming. Assisting him, Suharto learned the ways of *musyawarah* and *gotong-royong* in all their infinite variety. This experience also made Suharto deeply religious, another bond he shares with the peasants.

His warmth, simplicity, and understanding are what prompt people to discuss intimate and important problems even though they are not aware of who he is. In one village, when Suharto asked a farmer if he had any complaints the farmer said, "Yes, I do. Before I can sell my rice I have to pay the mayor a tax of two kilos of rice, and then when I take it to market I have to pay the soldiers on the road another tax in rice." As the conversation progressed, the farmer recognized him and began to apologize profusely—"I'm sorry, I'm sorry, Pak Harto, excuse me, excuse me." Suharto reassured him that this was exactly the kind of thing he, as president, needed to know. He ordered an immediate investigation and the corrupt practice was stopped.

Suharto's attention to detail uncovered a fertilizer racket. In one village, when he asked the farmers how much they were paying for fertilizer the price they told him was just half the official market price. He asked them how that could be. "We bought it from the village officials," they said. This aroused the president's suspicions, and as the trip progressed he checked government fertilizer storage barns. He discovered one large barn that was unguarded and from which officials of various villages had been stealing fertilizer. This incident prompted a complete reorganization of fertilizer distribution and led to the building of a new fertilizer plant to increase production as well as to the construction of a new plant to produce fertilizer bags.

Suharto refuses to spend the night in a hotel or government guest house. He prefers to be an unexpected guest in a farmer's home. Or, as happened late one afternoon when he was tired, he told the group to stop at a restaurant on the side of the road for coffee. After dozing in a chair at the table, he suddenly told his aides, "We'll spend the night here." In consternation, they pointed out that there were no beds. "It doesn't matter," he said. "We'll sleep on the floor." Which he and they did. Another night, in one of the poorest and most remote mountain villages, he slept in an open thatched hut and bathed with the villagers in the river as is their custom and only resource.

Sometimes he is not recognized at all. At one rice farm he sloshed his way through the irrigated rows of plants to question the farmer about his crop. The farmer told him he had increased his production on the last crop from two tons to three. "Do you use fertilizer?" Suharto asked. "Oh, yes, I do now," the farmer replied, and showed him what he was using and how. He answered the president's questions about prices and supplies and assured him he had no problems. Suharto asked how many children he had, and the farmer said he had thirteen. "Well, you must grow more rice next year, but don't grow any more children," the President said with a smile. "What you need is family planning." The farmer grinned, and as the group started to leave an aide asked, "Do you know who that was?" The farmer shook his head. "Who do you think it was?" The farmer, shrugging his shoulders, said, "Probably some tourist from Jakarta."

In such fashion the president has communicated with his people and learned from them. By the end of 1971 he had made five "silent inspection" trips throughout Java, each lasting a week, and trips to all the major islands. On a trip to Kalimantan, one of the big undeveloped islands, he visited a newly established village that is part of the government's "transmigration" policy in an effort to encourage Indonesians to move from crowded Java to the outer islands. There were five thousand new settlers, and when he asked them if they were happy, they told him no. He got to the heart of their needs, which were simple—140 little boats, one for each family, to take their produce downriver to market. He ordered the boats to be brought from a seaport the next day and to be paid for "from the pocket of the president," a special discretionary government fund. The settlers also needed a new water pump and wanted a mosque, both of which he paid for. As a result of this visit another five thousand settlers have emigrated from Java. The word spread quickly that Kalimantan must be a good place because the president cared enough to go there.

Suharto's deep concern for the people's welfare stems from the years in which he was a guerrilla army leader fighting for the nation's independence. "During the revolution, the people supplied us with rice and clothes and helped us in every way," General Bardosono comments. "We, in the army, felt that this was a loan. It wasn't on paper, but the president feels that he is paying back his debt to the people."

On a working trip to Irian Jaya, Indonesia's half of New Guinea, which comprises more than one fifth of Indonesia's land area, Suharto was greatly moved by the plight of its Stone Age aboriginals, numbering some four hundred thousand, who were living naked in the interior. Government and missionary groups have been working for many years to help civilize the tribes. Now, the need to prepare them for the encroachment of civilization is increasing, since the area has tremendous oil and mineral potential. Many contracts have been given and active development has begun. However, President Suharto's prime concern was that so many Indonesians were living outside society without even the minimum provision of food, clothing, and education. Government development plans call for social projects to be established ultimately, but the president felt that the Indonesian people could respond to this urgent need without waiting for government budgets to increase to the point where the problem could be tackled. He launched a nationwide foster-parent plan to help provide for Irian Jaya children of school age.

"I know that the rest of the people do not have much more money than the people there," the president told me, "but we must help our brothers in Irian Jaya. It does no good to try to teach the old ones, so we must work with the young ones and prepare them to live as Indonesians so they can help develop the country." A contribution of fifteen dollars enables a child to buy clothes, some tools like a shovel, knife, saw, or spade, and pay his school fee for a year. In the first program, foster parents have provided for over five thousand boys and girls selected from the villages to receive training.

President Suharto constantly seeks the involvement of all the people in the government's eonomic development programs. To increase development capital he initiated a National Savings Development Drive. He invited the whole population to take part and called on all parents and teachers to educate the schoolchildren to participate. "The drive is hoped to awaken our consciousness about the meaning of money to increase capital for development for our greater needs in the future," he said.

"The habit to be thrifty and to save is the life habit of a development-prone person."

Another of the president's personal campaigns is to urge farmers to increase animal and poultry production. He talks to them about protein values and recently ordered part of the state palace grounds to be converted to a poultry-farming exhibition. He has even started raising chickens in his own backyard.

Although development and domestic problems had the highest priority, during 1970 the New Order took an increasingly active part in international affairs and promoting international confidence in Indonesia's economic progress. The government took the initiative in holding the Jakarta Conference, attended by foreign ministers of Asian and Pacific countries, to help find solutions to the problems of restoring peace in Cambodia and stability and peace in Southeast Asia. Together with the member countries of the Association of Southeast Asian Countries, Indonesia established various joint economic, technical, and cultural projects. President Suharto attended the Summit Meeting of Non-Aligned Nations in Lusaka, Zambia. He also made state visits to Malaysia, Thailand, the Netherlands, the Federal Republic of Germany, and the United States.

On his trip to the United States President Suharto was particularly impressed with a Congressional Prayer Breakfast he attended and a trip to Disneyland. In explaining to congressmen how deeply religious Indonesia is, he told them, "Religion does not only free people from the backwardness of the past, it is not only a guidance and light in times of trouble, but religion can also accept all consequences of progress. In fact, religion should actually become a stimulus for mankind toward progress."

He told me that he was especially impressed with Disneyland because "before, I had known Americans only as businessmen. Disneyland is not only for children but for grown-ups. I saw it as a spiritual experience,where, in accordance with our Eastern attitude, I saw the close relationship and familiar approach between American children and their parents."

Suharto's own exemplary family life is also part of his rapport with the Indonesian people. During the revolution he married a former schoolmate. The Suhartos live quietly with their six children—three boys and three girls, ranging in age from twenty-two to seven—in a modest suburban home indistinguishable from adjoining houses except for guards posted in front.

President Suharto constantly equates the nation's life with family life. "Every man, every family, and any nation also, has to experience ups and downs," he told the people in his anniversary speech. "The important thing is our attitude. We must ever grow in maturity and in evaluating all those experiences. Let us make of this Republic of Indonesia a house for a big family, the Indonesian nation."

From the beginning, one of President Suharto's major goals had been to restore the power of government to the people by holding popular elections for a new House of People's Representatives, the legislative body of the state. The People's Consultative Congress is a superparliament, consisting of the members of the house plus regional and functional representatives. This assembly elects the president and vice-president for a five-year term and recommends such changes in the constitution and the structure of government as it considers advisable. However, under the constitution the president has supreme power in conducting the administration of the state, and his ministers are responsible only to him.

The Provisional People's Consultative Congress had agreed upon a general elections law calling for popular elections in July 1971. Many government and political advisors urged President Suharto to postpone these elections until the Five Year Plan had been given more time to be implemented. The president was adamantly opposed to any such delay. He considered general elections and achieving political stability a vital part of overall national development.

This was the first time the people had gone to the polls in sixteen years. There was great excitement throughout Indonesia, and nine rival political parties conducted vigorous election campaigns. President Suharto donated a new election van and 10 million rupiahs to each of the political parties. On election day, the village people put on their best clothes to vote. Out of 57 million registered voters, 53 million participated. The Golkar party, organized to support the Suharto government, won a major victory, receiving about 65 percent of the votes and commanding a clear majority in the house with 233 out of the 360 seats contested. The army had already received an allocation of 100 seats, as its personnel had been forbidden to participate in the election. The Nationalist party, founded by Sukarno, managed to win only 20 seats, and the Moslem Scholars party, a new political organization, emerged as the second main party with 58 seats.

A leading Indonesian businessman, who toured many villages and talked to voters on election day, told me that the reason the people voted for the government party could be summed up in one old farmer's comment. "I voted for Golkar," the farmer said, "because now for the first time in many years I can let my coconuts and bananas ripen on the trees and my sweet potatoes ripen in the ground. I don't have to cut my corn too soon and I can leave my clothes on the washline. I no longer have to worry about thieves. When I take my produce to market, I no longer am stopped by the army and made to pay a toll to travel the roads."

The first of the two basic short-term tasks that President Suharto had set for the nation in 1970 had been achieved: the general elections were a success. The second task—the successful implementation of the Five Year Development plan—was making steady progress. Most importantly, the country had reached a point where the results of many basic decisions and policies made in the early days of the New Order were bearing fruit on a scale that would further assure the development plan's success.

7
A Vast New Oil Province

There is a rule of thumb in the oil industry that the barrel of oil you use today someone had to start looking for five years ago. But that applies to the successful finds. The art and science of oil exploration still results primarily in the drilling of dry holes. On an average only one out of every nine exploration wells finds oil. The odds of finding a big commercial field are probably no more than one out of fifty. Although technology can land men on the moon, it has not yet found any means of directly locating oil in the earth. Sophisticated techniques have been developed to locate buried rock layers so shaped that they conceivably could form traps for oil, but whether or not they contain oil or gas can be determined only by drilling a hole.

It is a costly gamble. Onshore wells, depending upon their depth and location, can cost anywhere from $50,000 to $150,000 each. Offshore, the stakes are higher, ranging from $500,000 to $2 million or more per well. In any new

President Suharto of Indonesia (left) and Lieutenant General Dr. Ibnu Sutowo, President Director of Pertamina, with Robert O. Anderson, Chairman of the Board, Atlantic Richfield Company, at the September 1971 ceremony inaugurating the first Ardjuna offshore oil-producing platform.

unexplored area finding the first oil is the hardest, like trying to get the first olive out of the bottle. Pan American had walked away from Indonesia after more than twenty dry holes and a loss of $28,250,000. Ironically, Caltex subsequently contracted for some of the same Sumatran acreage and discovered oil. This, too, is a rule of thumb: dry holes frequently prove nothing. Oil finding requires persistence, faith, money and the willingness to take a chance. Another essential ingredient, as in the turn of the cards or the roll of the dice, is luck.

At first, luck did not seem to be with Pertamina and IIAPCO in Indonesia's first exploration in the Java Sea, off the coast of north Java, on its first production-sharing contract. Sinclair Oil Company (later merged with Atlantic Richfield) had acquired a controlling interest in the contract and was the operator. After a year of marine seismographic work to locate potential structures, the first well, drilled from September to October 1968 to almost 4,000 feet, was plugged and abandoned–a dry hole. The "exploration fever" thermometer began to rise when there was a show of gas in the second well but not enough to make it commercial. The thermometer rose a few more degrees when there was a small show of oil in the third test, but it was not commercial, either. However, the fourth well was another dry hole. At this point Lady Luck seemed only to tease, and the explorationists worriedly studied their cards as their pile of chips diminished. As the year ended they began drilling the fifth well. On February 4, 1969, the mercury exploded through the top of the fever thermometer. At 4,649 feet, low-sulfur crude oil was struck which tested at a flow rate of 2,600 barrels a day. Indonesia had its first offshore oil, which had been waiting 20 million years to be discovered. The offshore oil boom had begun. When two months later Sinclair-IIAPCO discovered a second big field there was great excitement.

Even so, the olives were not coming out of the bottle easily. By the end of the year Japex had drilled nine dry holes off the shore of northeast Sumatra. Union Carbide also had drilled nine dry holes, off the shore of Kalimantan, and Union Oil of

California one dry hole off the shore of northwest Sumatra. The majority of the other foreign contractors were still carrying out marine seismographic and airborne magnetometric exploration. At the same time, Pertamina was signing new production-sharing contracts, bringing the total to thirty-eight by the end of 1970. As new exploration drilling got under way in 1970 the dry hole score began to mount, but there were three bright spots. Sinclair-IIAPCO found a third big field. IIAPCO, which had taken a second production-sharing contract in the Java Sea off the shore of southeast Sumatra, struck oil after eight dry holes, opening up a new area. Unlucky Japex had pooled part of its off-shore Kalimantan contract area with Union, and Union struck the biggest well found up to that time, testing 11,300 barrels a day.

Both IIAPCO and Union started January 1971 with two new offshore discoveries, and in March, Mobil found oil onshore in north Sumatra. But by the end of November, after eight months of no major new discoveries, news reports were speculating that the much-touted Indonesian oil boom was dwindling. *Newsweek* reported that "prospecting has been so poor that several $10-million oil rigs lured to Indonesia from the busy Gulf of Mexico now sit idle in port. Only a handful of wildcat holes are being drilled, and many small companies are strapped for cash."

Because of the risky nature of the game oil explorers have their doubting periods, but nothing is better calculated to rekindle a lagging boom than new discoveries. Beginning in December 1971 and through mid-1973, scarcely a month passed without one or more important discoveries being announced in areas covered currently by forty two production-sharing contracts. After a slow start, by mid-1973 56 new oil and gas fields, of which 35 were onshore and 21 offshore, had been discovered since General Ibnu had taken over guidance of the nation's oil future in 1966. Of these, 6 had been discovered onshore by Pertamina as the result of its own exploration activities.

There was now no doubt that Indonesia was a vast new oil province. "This is only the beginning," Robert Martin of Caltex, shareholders' representative for Texaco and Standard Oil of California and a veteran international geologist, told me:

> The potential for the discovery of oil throughout this three-thousand-island archipelago makes it one of the largest unexplored oil areas in the world. Of course, everyone is hoping to find another billion-barrel oil field like Caltex's Minas field in Sumatra; however, there is such a variety of geological areas in this great area that you could find another Middle East, get out of it, and then get back into another one again. There will be big fields and marginal fields and room for lots of Oklahomas and Kansases. This is going to be a long-term development process. Undoubtedly acreage will change hands many times and new methods will come into play. An added exploration incentive is that this will probably be mostly low-sulfur crude.

The significance of the discoveries already made is that they are not concentrated in one area but are scattered throughout the 965,000 square miles, onshore and offshore, which are under production-sharing contracts. The excitement of such widespread exploration is that each exploration well stands the chance of discovering a new oil or gas target, which in turn stimulates further exploration.

There had never been any significant gas production in Indonesia, but in December 1971 Mobil struck gas on its onshore contract area in north Sumatra. The announcement of the discovery did not attract great attention; however, subsequent development drilling confirmed the discovery of what promises to be one of the world's big gas fields. Plans are being made for the investment of a billion dollars in a natural gas liquefaction plant and liquefied natural gas (LNG) tankers. Another big gas discovery was made in 1972 by Huffco, an independent American group, on an onshore contract in east Kalimantan. Phillips and Cities Service have made noncommercial offshore gas discoveries in the South China Sea and the Java Sea. All this means that Indonesia is in the process of developing the capability of being a very large exporter of LNG, one of the most desirable sources of clean energy, to serve the Japanese and U.S. West Coast markets.

Another new target was discovered in early 1973 when Petromer Trend, an independent American group, brought in an onshore discovery well in remote Irian Jaya at the eastern end of the archipelago. This was a different type of discovery, as the oil was accumulated in a limestone reef rather than in sandstone formations. Productive reefs are difficult to discover, but they are prolific oil structures and have yielded a substantial part of the world's oil production. The discovery of one reef means that the region is likely to contain many others, as reefs are rarely if ever isolated. The first well tested at a flow rate of approximately 4,600 barrels a day, at a depth of about 3,500 feet. The third well came in at a spectacular 23,600 barrels of oil daily, Indonesia's largest well.

It is still too early to evaluate how many billions of barrels of oil reserves the new discoveries will eventually prove through development, but Indonesia's production has more than doubled since 1966, reaching 1.3 million barrels daily in mid-1973, and 1.5 million barrels daily is expected by the end of the year. Production is estimated to reach approximately 2.5 million barrels daily in 1976 and almost 3.0 million in 1979, just on the basis of present discoveries. Production from new discoveries, resulting from exploration drilling already under way and yet to come, could greatly increase these estimates. Now that the vast new oil province has been proven, the initial exploration commitments of approximately half a billion dollars for the forty-two production-sharing contracts signed by mid-1973 will multiply many times as new discoveries are made, encouraging further risk capital. Although 965,000 square miles on water and land have been contracted, many more thousands of square miles of promising territory have not yet been opened for applications, and more will become available as contractors make their required acreage relinquishments over the next few years. Consequently, Indonesia will see another vigorous round of competitive bidding and exploration in the near future.

Historically, the most formidable barriers to full development of the earth's petroleum resources have been the restrictions placed on exploratory activity by government policies. This is the effect upon physical geography of "human geography"—the attitude of man toward his mineral resources. The removal of such barriers in Indonesia by the principle of production-sharing contracts not only is resulting in full development of Indonesia's petroleum resources but has created a climate of cooperation and understanding between foreign contractors and Pertamina which is unparalleled in the history of relationships between international oil companies and host governments.

The problem of management control, to which the large international oil companies initially objected so strenuously, has turned out to be no problem at all. Production-sharing contracts create a working partnership in which Pertamina is the senior partner. Pertamina approves the contractors' work proposals and all concerned have a common interest in making right decisions. Pertamina's Division for Coordination with Foreign Contractors handles all matters for contractors concerning government relations as well as such things as customs requirements, work permits, and compliance with safety regulations. This leaves the contractors free to concentrate on finding oil and developing oil production. Furthermore, it removes them from the political scene and makes their role more acceptable to the public. Charges of political domination by foreign governments have been a major source of trouble and friction for oil companies in their international operations, and mounting public pressures have in some cases led to nationalization of the companies' properties.

The real success of production sharing has been due to the spirit with which Pertamina and the contractors have carried out the contracts. "You always know where you stand with General Ibnu, and you can depend upon getting prompt decisions," one company official said, summing up the opinion of all. There is a no-nonsense, shirt-sleeve, working relationship between the government oil company and foreign contractor employees, and there is camaraderie rather than antagonism or suspicion between them. A Petroleum Club has been formed whose new facilities in downtown Jakarta are used as a center for joint social gatherings and technical lectures. The club, with four hundred members, also has a swimming pool and tennis courts. General Ibnu, an expert golfer, started the custom of having Pertamina and the foreign contractors take turns hosting monthly golf tournaments.

Further to meet the needs of the professional aspects of the industry, the Indonesian Petroleum Association was formed in 1971 to act as a sounding board between government and industry on such matters as environmental control, training, and education. In 1972 the government formed a pollution team to recommend pollution control regulations designed to prevent pollution from oil exploration and production activities. Financed by voluntary contributions from the Indonesian Petroleum Association, the study team visited various countries around the world to learn how such problems were being handled. In March 1973 the Ministry of Mines issued pollution control regulations for Pertamina and its foreign contractors and promulgated requirements for emergency plans to take care of any accidental oil spills.

"One of the reasons why the foreign oilmen and Pertamina get along so well is that ideologically they believe in the things upon which our constitution is based," General Ibnu explains. "They believe in democracy, an equitable and civilized humanity, and social justice, just as we do."

Although Pertamina's ultimate goal is to conduct all Indonesian oil activities once it has acquired the requisite trained technical personnel and can generate sufficient capital investment resources, General Ibnu has no dog-in-the-manger attitude. He makes economic decisions and sets policies that are based entirely on what is most profitable for the country today. "One dollar today is worth more to Indonesia than ten dollars will be ten years from now," he says.

One of the best illustrations of how he applies this is the way in which Pertamina has handled its relationship with Caltex. Caltex is the country's largest producer, with almost 77 percent of Indonesia's current daily production, and it reached a level of over one million barrels daily production in mid-1973. In 1968, Caltex wanted to acquire 1,600 square miles for exploration in central Sumatra, adjoining its production area, as an extension of the 60-40 profit-sharing Contract of Work under which it had been operating since 1963. Many oil companies were eager to make production-sharing contracts on this area, but General Ibnu, despite his objections to the principle of Contracts of Work, agreed to give Caltex the requested extension. After a thorough study of economic projections, he was satisfied that Indonesia would receive more profit by letting Caltex have the area under the old terms than by giving it to a newcomer under a production-sharing contract, as Caltex had established facilities that would reduce development costs.

Again, in 1971, General Ibnu characteristically set a new precedent that was the reverse of the trend in other countries toward starting de facto take-over of properties by national oil companies before concessions expire. Caltex's Contract of Work on 2.4 million acres in central Sumatra was due to expire in 1983, at which time its properties would revert to the nation. General Ibnu announced an agreement whereby Caltex will continue operations on its present properties for eighteen more years, until the year 2001, under a production-sharing contract for the extended period. At the same time, Pertamina signed a new production-sharing contract with Caltex for an additional 5.5 million acres adjoining the older holdings.

"If Pertamina had taken over in 1983, Caltex would have had to build new production lines and facilities for their new fields under the new contract, and if their costs had gone up Pertamina would have made less money," General Ibnu points out. "Also, in regard to the new contract, because they have all the physical facilities they can produce oil at less cost than a newcomer. It is popular in the world today for a country to own and operate everything, but if the companies can do it better, then it is better for a country to do it with them."

General Ibnu estimates that because of the huge capital requirements for exploration and production it would have taken Indonesia twenty or more years to try to develop its oil resources by itself. There was an additional economic factor to be considered. Foreign oil contractors could justify costly dry holes because they are part of the risky nature of their business. However, a national oil company spends the people's money. There is no justification for a developing country with inadequate capital resources to spend millions in costly dry holes when roads, housing, schools, and other basic necessities are urgently needed and when private capital is available for such risks.

This is another reason for Pertamina's policy of letting private oil companies assume the high risks of exploration offshore and in remote, undeveloped onshore areas. Pertamina has largely confined its own exploration to onshore areas adjacent to old, established production in Sumatra, Java, Kalimantan, and Irian Jaya. This has enabled its technical personnel to increase their experience with advanced technology in exploration and production at relatively low risk-cost and has already resulted in one bonanza as well as five other discovery fields. Royal Dutch-Shell had carried out extensive geological, geophysical, and drilling explorations on the north coast of west Java and abandoned the area as unpromising. When Pertamina acquired the area, it contracted for new seismographic surveys and reevaluated its possibilities. In 1969 a third exploration well made a major oil discovery, the Jatibarang field. Production from this field, which is being exported to Japan, began in 1973 and by 1974 will double Pertamina's pre-Jatibarang production of 100,000 barrels a day. Gas from Jatibarang will be used as feedstock for a plant to manufacture urea fertilizer, which Indonesia needs in increasing quantities.

In early 1970 General Ibnu concluded negotiations with Stanvac for the purchase of its Sungei Gerong refinery in south Sumatra, which has a daily capacity of 75,000 barrels for $4,750,000. To enable Pertamina to integrate this refinery with Pertamina's 110,000-barrel-a-day Plaju refinery directly across the Musi River, Stanvac made an interest-free loan of $2 million, with the loan and the purchase price to be liquidated within six years. At the same time Pertamina gave Stanvac a large extension to its work contract areas for new exploration. The refinery purchase placed all national refining facilities under Pertamina. Pertamina had also been busy expanding them; it completed a new 50,000-barrel-a-day refinery at Sungei Pakning in south Sumatra at the end of 1969 and a 100,000-barrel-a-day refinery at Dumai in central Sumatra in late 1971, bringing Indonesia's total daily refining capacity to 411,000 barrels. Another 100,000-barrel-a-day refinery is planned, to be constructed within the next few years, to meet rapidly rising domestic requirements and provide additional export products. Pertamina now markets petroleum products in eleven foreign countries.

Indonesia's fast-growing economy has been increasing its consumption of petroleum products at the rate of 9 to 10 percent yearly, which exceeds the worldwide average annual growth rate of only 6 percent. Domestic consumption increased by 35 percent between 1966 and 1972 and is estimated to increase an additional 39 percent by 1975 and double by 1980. Pertamina handles all domestic marketing operations and during the past five years has built or completely modernized more than two hundred service stations, some of them attended by pretty girls in smart uniforms.

Pertamina markets about one fourth of all crude oil exported from Indonesia, which includes part of its own production, the amounts it chooses to market from its share of production-sharing contracts, and most of the 20 percent in kind to which the government is entitled from production under Caltex and Stanvac Contracts of Work. Pertamina's ability to develop crude oil export markets in more than six countries is one of the keys to its success as a national oil company. Furthermore, it has increased government revenue from its profit-sharing Contracts of Work with other oil companies.

Under the 1963 Contracts of Work, Indonesia's 60 percent share of profits is based on the prices the big oil companies realize in the market. If the government did not elect to take any of its 20 percent option on gross crude production and attempt to market it, the government would be dependent upon the oil companies' determination of realized prices. In 1964 Caltex was selling crude from its Duri field in Sumatra at an export price of $1.49 F.O.B. per barrel. Pertamin, the one of the three then-existing state oil companies with which Caltex had its work contract, found a market for the same oil at $1.63 a barrel. When the government advised Caltex that it would take up part of its option, to sell at this price, Caltex contended that it was overpricing and would ruin the market. However, Pertamin was able to sign a two-year contract with a Japanese buyer at this price and made Caltex an offer to buy all its Duri crude. Caltex promptly announced a change to $1.63 in determination of "realized prices" to the government, proving the benefit to the government of being able to compete in the open market.

To handle the great increase in its exports and in its domestic distribution throughout the islands, Pertamina has developed the largest tanker fleet in Southeast Asia and constructed thirty-five sea terminal depots. At the end of 1972 Pertamina operated a fleet of 104 interisland and oceangoing tankers, of which 49 were Pertamina-owned. The fleet has a deadweight tonnage (DWT) of almost 1.4 million tons, at least five times that of the 1966 fleet. Pertamina is adding a number of large, 100,000-DWT tankers to its fleet. Besides using its oceangoing tankers for its own exports it charters them to others in the international tanker market to increase Indonesia's foreign exchange earnings.

Pertamina is laying the foundation of a petrochemical industry that will be of great importance to the country by supplying it with synthetic materials for manufacturing. A $14-million polymerization plant completed in mid-1973 in south Sumatra is producing at the annual rate of 20,000 tons of polypropylene resins for the manufacture of plastic household goods and synthetic fiber from which knitted goods can be made.

The United Nations Industrial Development Organization has completed the first phase of a study of Indonesia's potential petrochemical production, with special reference to the use of natural gas as a feedstock. The study foresees the possibility of investment needs in excess of $200 million. The World Bank also is making a comprehensive survey of Indonesia's petrochemical future.

General Ibnu recently sponsored Indonesia's first petrochemical seminar with twenty expert guest-lecturers from the United States and Great Britain. "We have already met the required conditions for developing a petrochemical industry," General Ibnu told the seminar. "We have a stable oil industry, favorable geographic location and rich natural resources for both domestic and export markets, the ability to import technical know-how and technology, and the favorable attitude of the government." He arranged the seminar, which was open also to other government agencies and university faculties, for Pertamina senior officials "in order to change their old views about the petrochemical industry into new ones."

In addition to sponsoring such special events, as Pertamina has expanded General Ibnu has intensified his original insistence on the importance of technical training. Pertamina employees in all phases of management and operations are sent abroad in increasing numbers to take special courses in the United States and Europe.

Pertamina's growth since 1968 had made it appear to be the fiery head of a comet streaking through the sky, with other government agencies and private companies forming the tail. The objections to such speed, which had provoked the public criticisms and attacks charging corruption in 1969 and early 1970, were being answered in exactly the way General Ibnu had told his staff they would be. "We will have our headlines," he said at the time, "but they will be about growing production, new refineries, and improved domestic supply. Those headlines will let the people know where the money goes. Only results answer such charges."

Pertamina had formed such joint ventures as the Far East Oil Trading Company to seek outlets for crude oil in Japan and the Hong Kong Tugu Insurance Company to handle Pertamina's and foreign contractors' insurance. It had also created such subsidiaries as Ocean Petrol to Pertamina's international shipping fleet, Electronika Nusantara to handle telecommunications, and Pelita to handle aviation services for Pertamina and foreign contractors. This had all been done before production increases and exploration successes in production-sharing-contract areas had become evident on a large scale. General Ibnu had been planning Pertamina's expansion to handle this projected growth. Public suspicions of Pertamina's motives in forming these companies were based on lack of information and on circulation of misinformation. Furthermore, the very size of Pertamina led to claims that it was beyond government control. Its aggressiveness in obtaining financing for its expansion from international banking sources and borrowing from companies against future crude-oil production had alarmed even some of the worldwide government banking circles that were supervising Indonesia's overall government development loans—loans that were essential to the nation's continued growth and development.

President Suharto had publicly defended Pertamina's expansion and contributions to the nation, on the basis of a study by a presidential commission, and had directed the ministers of mines and finance, the National Development Planning Council, and experts from other ministries to draft a law specifically to define and regulate Pertamina's activities. The law, which the parliament passed and which became effective on January 1, 1972, cleared the air and removed the basis for claims that Pertamina was a law unto itself. Pertamina welcomed clarification of the new law in order to continue the planned expansion of the country's oil and gas resources.

The Pertamina Law continued the jurisdiction of the Ministry of Mines over Pertamina but created a Government Council of Commissioners to approve Pertamina's policies, budget, borrowings, establishment of subsidiaries and joint ventures, and various other activities. It provided for expansion of operations beyond oil and natural gas only with the consent of the president and if such projects are connected with oil and gas operations. The council consists of the minister of mines as chairman, the minister of finance, and the chairman of the

National Development Planning Commission. Provision was made for the addition of two more members at the discretion of the president, who retains final authority on all decisions. The law specifically defined the method by which Pertamina's operations would provide the government with its revenues from the country's oil and gas resources. The law also specified that Pertamina would be managed by a Board of Managing Directors consisting of the president-director and five others, all to be appointed by the president. In addition to General Ibnu as president-director, President Suharto named as directors the heads of Pertamina's five principal divisions, most of whom were veterans of the state oil enterprises.

The legal clarification of Pertamina's role in the nation's development also served to focus full attention on what its contributions to the country had been in such a remarkably short time. In 1966, income from oil to the government was only a little over 5 percent of its total. In 1972, oil earnings contributed almost 40 percent of the government's budget, up from 33 percent the year before. By the end of 1973 oil will contribute an estimated 50 percent. In value, oil export revenues grew by almost 65 percent in 1971, to approximately $1 billion in 1972, for the first time exceeding the total of all other exports and accounting for over half of the country's gross foreign exchange earnings. This $1 billion is in contrast to oil export revenues of $217 million in 1966. In 1972 the petroleum sector grew almost five times more rapidly than did the economy as a whole. More important, oil exploration and development expenditures, a barometer of future growth, are expected to grow by two thirds during 1973, from $271 million in 1972 to a projected $455 million.

At long last, the goal set forth in Indonesia's Constitution of 1945, that its natural riches be developed "for the greatest benefit of the people," is being realized. Development of the oil and gas riches of its land and waters is giving Indonesia the means to reap other harvests. Furthermore, the opening of the vast new oil province is making other contributions that cannot be expressed statistically but whose effects are multiplying in unusual ways.

8
"Develop the People"

The world's preoccupation with international development during the past quarter century has been on a scale new to history. There are as many different interpretations of what constitutes effective "development" as there are domestic or foreign economists, sociologists, engineers, bankers, businessmen, politicians, and experts in other fields working in any given developing country. However, there is one point upon which all experts agree: development must come from within. It is a question of national will.

It was quite clear to General Ibnu that Pertamina's role was to strengthen and expand Indonesia's national will. "To catch up in this world of modern technology, the first thing is to develop the people," he said. "Development begins with the development of people."

General Ibnu's concept of how to develop the people was an innovative, unorthodox mixture of the principles he had successfully applied to Pertamina's growth combined with new ideas that its expansion made possible. To him, Pertamina's responsibility was not just to generate financial revenue for the government from the development of oil resources but also to be a catalyst in the overall development of the nation's human resources. From the beginning he had emphasized the importance of training as many Indonesians as rapidly as possible to handle every phase of oil operations. Now, he channeled Pertamina's know-how and experience into becoming a driving force in modernizing Indonesia and creating new opportunities for developing the people. Pertamina made itself valuable by profiting the nation in ways that cannot be reflected in a commercial accountant's profit and loss statement.

"If we were just an oil company trying to make a profit for our shareholders, Pertamina would not be doing a lot of the things we are doing," General Ibnu told me. "However, our shareholders are all the people of Indonesia, and we are a development company for the nation. Pertamina must help bring the country into the modern economic mainstream. Because Pertamina is my country's biggest commercial enterprise and has had the most experience, we must take the lead."

It was this approach that had provoked much of the criticism and controversy over many of Pertamina's varied activities from those with less imagination and insight. However, it resulted in unexpected sources of revenues for the government and accelerated the development of new enterprises that, in turn, provided opportunities to train more Indonesians in new skills.

As I visited Pertamina's operations throughout the islands, it was obvious that the only thing lacking, wherever I went, were signs saying, "Your Oil Rupiah at Work." Pertamina has a total involvement in contributing to community development wherever it operates. Since its operations, which include exploration, transportation, refining, and distribution of products, are so far-flung among the islands, its presence is felt everywhere. Unlike a commercial enterprise, Pertamina has undertaken people-development activities not because it has to or because they represent calculated good business practice from a public relations viewpoint. It has done so because of its philosophy of leadership responsibility. Pertamina employs more people than any other single company in Indonesia, and General Ibnu's quest for excellence and self-reliance has imbued its forty thousand officials, technicians, and workers with an enthusiasm and esprit de corps seldom found in any big organization.

A young field engineer in north Sumatra typified this feeling when he told me, "My father was a soldier in the revolution, and he told me that I had to work for my country. I studied petroleum engineering so I could work for the national oil company. Even if a foreign contractor offered me more money, I would not work for them. I only work for my country. As a revolutionary, my father was of the generation of 1945. We young people now have a chance to carry out the revolution, and we call ourselves the generation of 1966."

I found this same dedication among the students at the Academy of Oil and Gas at Cepu in central Java, which now offers a regular three-year, college-level course in every branch of the oil industry. The students have excellent laboratory and classroom facilities, but their real classrooms are an actual oil field and refinery where they "work while learning and learn while working."

Before they are given an opportunity to attend the academy, the students must work in some capacity for the oil industry for a year to demonstrate their seriousness about their choice of profession. Character-building also is part of the training. When they return to Pertamina after graduation their job responsibilities are frequently changed to see how they respond to discipline. However, they know that there is no limit to how high they can go in the company.

The academy continues to offer special short courses, but today these are technical and refresher courses for post-graduates. The academy also holds orientation courses on the petroleum industry for the armed forces, to develop a corps of oil specialists, and for government officials in various departments who deal with the oil industry in their daily work.

The original "mental upgrading" courses had such a profound effect upon Pertamina's employees that the study of philosophy continues to be as important to them as the study of technical manuals. When I visited the Prabumulih oil fields in south Sumatra, a former Communist labor union stronghold, the management staff showed me two volumes, with charts, which they had compiled for oil-field workers to use in study and discussion groups. The first volume was called "Oil Field Men's Way of Life." Based on General Ibnu's doctrine of "Work while you learn and learn while you work," it stressed the importance of a "high Frontier Spirit for oil workers," and detailed their three main tasks, consisting of professional duties, nonprofessional duties, and spiritual and mental guidance duties. The second volume was called "Productivity Determines the Standard of Living." It explained the philosophy of management and the importance to the individual of increasing his own productivity.

The volumes were a spirited, rare mixture of psychology, self-help, religion, philosophy, economics, and sociology. They also provided a course in citizenship, community development, and patriotism.

Pertamina's emphasis on community development takes many forms. It has given priority to the building of roads and bridges which, in addition to their oil-industry uses, have opened up new farm-to-market outlets. Pertamina oil-camp power plants supply villages with electricity, and the camps invite the villagers to use their waterlines. At its camps Pertamina also shows weekly movies, which are attended by as many as twenty-five hundred people from surrounding villages. "We alternate their favorite cowboy movies with documentaries we get from various embassies," a Pertamina public relations man explained, "as we want to teach the people about the world."

Palembang, in central Sumatra, where the Sungei Gerong refinery is located, traditionally was bypassed by big events because of its remoteness. So Pertamina built a large sports stadium that President Suharto inaugurated for one of the national annual student games and for competitive sports teams from all over Indonesia. Pertamina also gave Palembang a town clock, a handsome minaret for its mosque, and a television station. It also provided a television station for Medan, the largest city in north Sumatra.

Pertamina is a strong supporter of national contractors in order to strengthen the development of private enterprise. For example, for the distribution of petroleum products it gives new trucks to contractors under a five-year-loan plan. At the expiration of the loan period the contractor may buy the trucks on an installment basis.

General Ibnu believes that setting high standards is an essential part of developing the people. Material conditions as well as mental attitudes must be upgraded to provide incentives and demonstrate to the people that poverty is not inevitable or a matter of fate, and that productivity pays off in material benefits. Consequently, Pertamina pays its employees salaries that are above the national average and gives them many fringe benefits. In the oil camps, it provides modern housing and recreational facilities that would do credit to any American suburb. The lowest-ranking laborers are provided with attractive two-bedroom, double housing units with gardens. A typical medium-class employee's home I visited in north Sumatra was a spacious, three-bedroom ranch-style house complete with television and displaying golf trophies won by both husband and wife in company tournaments.

"If by our example we show people that there is another way of life, and that it can be achieved by Indonesians, then we are showing the people the benefits of modernizing our economy," General Ibnu says. "From the bottom to the top we must demonstrate in all our operations that we are running a modern and prosperous business. It enables us to negotiate as equals with companies and governments worldwide and obtain the best possible terms for Indonesia. This not only is practical from a business viewpoint but also sets an example for other sectors of the Indonesian economy."

This philosophy contributes greatly to the pride employees have in working for Pertamina and stimulates participation in community action. General Ibnu's wife, Sali, has been a leader in organizing women's clubs and Boy and Girl Scouts in all the camps. The women's clubs actively promote family planning and welfare projects in the villages, as well as educational, cultural, and social events.

Although General Ibnu stresses the development of technological skills and material benefits, he believes that a "soul and living unbalance" is created if technology and science are not accompanied by the opportunity for religious development. Indonesians believe deeply in God and that man is a spiritual being. Although this belief takes many mystical forms, the society is predominantly Islamic. General Ibnu, who has twice made the pilgrimage to Mecca, considers Pertamina's support and aid to religious activities to be one of the company's obligations. It has built mosques for camps and villages, established religious schools and libraries, and sponsored religious lectures and publications. Its support is extended to Christian groups as well.

Another remarkable innovation is General Ibnu's policy of providing modern, up-to-date services and facilities for the foreign contractors. In developing countries throughout the world, foreign oil, mining, industrial, and agricultural companies traditionally have been forced to allocate a great part of their investment costs to providing and building everything necessary for their operations simply because adequate facilities did not exist. The host government was host in name only. Developing countries have benefited greatly by foreign development of the infrastructure and building of facilities of all kinds for foreign and domestic personnel. Direct private foreign investment, of which almost half the cumulative total worldwide is in the extractive industries, has added substantially to the real national income of developing countries. This was the case in Indonesia, of course, with the investments of foreign oil companies under the concession system and Contracts of Work. However, with the introduction of the concept of production sharing to the nation a new situation existed.

The sensitive political issues of ownership of national resources had been settled. The policy of the New Order is to encourage foreign investment, and this is being achieved on a major scale in the oil sector. Under production-sharing contracts foreign oil companies in increasing numbers were making a direct contribution by assuming the high risks of exploration, principally offshore, and their discoveries were multiplying. However, from his analysis of the situation General Ibnu concluded that the Indonesian economy was not benefiting as much from the secondary effects of the expanding oil industry as it could. The foreign contractors were spending very little money in Indonesia. Drilling equipment and supplies were purchased abroad. Most of the foreign personnel and their families headquartered in Singapore where there were better living, office, medical, and recreational facilities than in Indonesia. General Ibnu, with the enthusiastic approval of President Suharto and his economic team, decided to capture this economic growth for Indonesia and make it a host country in fact as well as name.

His vigorous, imaginative program to achieve this objective has resulted in astounding changes in Jakarta and other parts of Indonesia. It has increased employment and the establishment of new businesses and services. It has increased the efficiency of oil operations as a whole and greatly expanded Pertamina's capabilities as a development company for the nation.

There is an acute housing shortage in Jakarta, so Pertamina is constructing a twenty-story Oil Tower to provide modern apartments for foreign oil contractors and their families. The tower also has a department store, conference room, bowling alley, restaurant, swimming pool, and parking areas. In addition, Pertamina is constructing Indonesia's first planned, modern housing complex of one hundred luxury homes on forty-five landscaped acres in the Jakarta suburbs. The recreational area will include a large swimming pool, tennis courts, children's playgrounds, bowling alleys, and a health club with sauna baths. There will be a community shopping center, theater, meeting

Pertamina has built this new Oil Center office building in Jakarta to accommodate foreign contractors who were formerly scattered throughout the city.

hall, and clubhouse with guest and party rooms. These homes, costing $100,000 or more, will be rented to foreign contractors. Pertamina also gave a land grant to the Joint Embassy School for a school for children of diplomats and foreign contractors.

Hospital facilities were so inadequate in Jakarta that foreigners routinely went to Singapore for medical treatment. Pertamina has built a three-hundred-bed, seven-story hospital that is one of the most modern and best equipped in Southeast Asia. In addition to providing service for foreign contractors and the diplomatic corps, it is open to the general public. The hospital also provides a greatly needed center for advanced training of medical students and medical personnel.

In west Java, at Chirebon, where a Japanese construction firm built export terminal facilities and a sea-loading line for Pertamina's Jatibarang discovery field, Pertamina built an attractive, artistic complex of guest houses with a restaurant and golf course, initially to accommodate the Japanese contractors. A swimming pool has been added and the guest houses have been dedicated as a motel to encourage tourism. This is a prototype for other motels yet to be built. Pertamina is also helping upgrade existing hotels in various parts of Indonesia. General Ibnu believes that through its international activities Pertamina can help develop tourism, another excellent source of national revenue. For this purpose Pertamina and a group of American oil companies opened the popular and colorful Ramayana Restaurant in New York City, featuring Indonesian food, dances, and music.

In Jakarta's business area, Pertamina built a $6.5-million, eight-story Oil Center office building so that foreign contractors can avoid the inefficiency of having their offices scattered all over the city and to encourage oil-field service companies to move their headquarters from Singapore to Indonesia. It also set up a $6 million telecommunications system to service the oil industry so that Pertamina and the contractors can communicate worldwide by telex. This system provides for twenty-four-hour, microwave voice communication with offshore platform rigs and remote field operations; emergency air-to-ground communication by helicopters; facsimile transmission of charts, diagrams, and photos; and data transmission, if required, for computer operations. The completely integrated telecommunications network also serves civilian needs in many areas. Pertamina made it possible for Indonesian television viewers to see direct telecasts of the 1972 Olympics in Munich.

Pertamina has provided the Indonesian Petroleum Institute, the scientific arm of the Ministry of Mines, Oil and Gas Affairs Department, with modern laboratories to provide service to Pertamina and foreign contractors, in crude-oil analysis, product testing, core analysis, and micropaleontology. The institute also has a staff of specialists who work on exploration, production, refining, and pollution studies.

Pelita, Pertamina's wholly owned aviation subsidiary, is a nonprofit company organized solely to serve Pertamina and to provide aviation service at cost to foreign contractors. In addition to fixed-wing and jet aircraft, Pertamina acquired twenty-four helicopters to transport personnel and equipment to remote areas where there are no adequate roads. Pertamina has given helicopters and fixed-wing aircraft to the Ministry of Defense and Ministry of Communications which, in turn, train pilots for Pelita.

At Batam Island, off the shore of central Sumatra and directly across from Singapore, Pertamina has established a supply base for contractors servicing their offshore drilling operations. Pertamina, together with Nissho-Iwai, a major Japanese trading company, and Pacific Bechtel, an American worldwide engineering and construction firm, formed a multinational joint venture and prepared a master plan for a $372-million development of the island. The ambitious project includes petroleum and general cargo marine terminals, a road network, a petroleum refinery, steel fabrication shops to support the large-scale petroleum activities on and off the shore of Indonesia, a free port for specialized manufacturing plants, utilities, and a livestock industry. There will be a planned industrial community complete with housing, educational, recreational, and health facilities. Part of the island will also be developed for tourism, to lure vacationers from Singapore.

The remarkable thing about all these developmental achievements is that they have been carried out in a period of less than five years. At the same time, Pertamina has steadily developed as an integrated oil company. All of its activities have complemented each other, strengthening its oil operations as well as providing a catalyst for the nation's development in so many different ways. National oil companies in the majority of oil-producing countries are, in effect, subsidized by their governments, as they have not yet learned how to operate efficiently or profitably. Pertamina is an outstanding exception. It has been a do-it-yourself company from its inception. It has achieved its victories by the work it has done. It has competed successfully in the world marketplace and engendered the confidence of the international private banking circles that have made multimillion-dollar loans to finance its expansion.

Despite this unusual record, Pertamina's critics, including some international bankers who participate in development assistance to the Indonesian government, continue to find it difficult to understand why the government approves Pertamina's carrying out so many varied development projects, particularly those related to providing services and facilities for the foreign contractors.

When President Suharto inaugurated Pertamina Central Hospital in Jakarta in 1972, he explained the government's viewpoint very simply. He pointed out that because state funds were limited, the government had never given Pertamina any money for its operations but, on the contrary, had charged it with meeting the fuel needs of the people, boosting oil production to earn the biggest possible amount of foreign exchange for the purchase of capital goods and raw materials, and carrying out development activities.

"If foreign contractors are obligated to co-ensure production, then service to them also should be ensured," President Suharto said. "If we fail to do so, then the situation would be that, on the one hand, the foreign contractors are obligated to earn as much foreign exchange as possible while, on the other hand, they have to be allowed to use the foreign exchange, for instance for medical treatment abroad or for staying abroad due to the lack of facilities here. For this reason, Pertamina has been justified in building modern office buildings, for instance, and also this modern hospital." He emphasized that such development projects built by Pertamina "have wide strategic objectives and are very important for development purposes."

Building a psychological infrastructure is, in some respects, as important as building a physical one. It is with this in mind that both the government and General Ibnu believe that Pertamina's catalytic role will spark private entrepreneurship to generate more facilities and services and further expand the national will for development.

9
Sowing and Reaping

An ancient Indonesian proverb says that he who gives rice for planting is greater than he who gives rice for eating, and that he who teaches how to plant and grow more rice is greater than either of the others.

In its new concepts of teaching Indonesians "how to plant and grow," the New Order is greatly expanding the meaning of Indonesia's traditional symbol of prosperity–*Dewi Sri,* the goddess of rice. It has concentrated literally on planting and growing more rice as the fundamental basis of its development program and at the same time created incentives for the development of *all* its resources.

When the New Order began putting Indonesia's house in order, agriculture contributed nearly 60 percent of its national income, employed almost 70 percent of the population, and earned nearly 70 percent of its foreign exchange. Consequently, agriculture was the most critical area for national income growth. By its very nature, even marginal investments in

"He who gives rice for planting, is greater than he who gives rice for eating, and he who teaches how to plant and grow rice is greater than either."

agriculture produce immediate results from which the people benefit. The New Order devoted the majority of its foreign aid assistance and new oil revenues to agricultural development and rehabilitating the infrastructure, with priority to servicing agriculture's needs. This, in turn, triggered development in the industrial sector and set off the chain-reaction of encouraging more foreign and domestic private investment in all sectors, which increased the confidence of foreign governments and resulted in greater aid assistance pledges.

This commonsensible formula is working so well that Indonesia is exciting more interest than any other less-developed country. Indonesia is so busy developing that the development story would have to be projected on a 360-degree motion picture screen to convey the multiplicity, variety, and feeling of all that is taking place.

The wide spectrum of development is also giving Indonesia the capacity to cope with setbacks when they occur. Early agricultural successes prompted the government to predict that it would achieve its goal of self-sufficiency in rice by 1974. The government launched a "mass guidance" system for farmers which provides them with intensive extension courses to improve their skill and know-how, a better credit system, distribution of fertilizer and improved seeds, better marketing and processing facilities, irrigation systems, and a favorable floor price for rice as an incentive. Consequently, rice production, which is half of the agricultural output, was increasing by 6 to 7 percent in 1970 and 1971, exceeding the Five Year Development Plan's targets. However, in 1972 the rice goddess was fickle. There was a serious drought, and rice production declined by as much as 8 percent. Other crops suffered as well. The government had to import approximately 1.5 million tons of rice, three times the amount imported the year before.

Such an event would have created a crisis a few years earlier. There was, indeed, a sharp increase in rice and other food prices, but there was no consumer panic or increased prices other than food. The government was able to finance half of the rice imports from aid sources and to buy the rest with $90

President Suharto chatting with a worker at the Pusri urea fertilizer plant in south Sumatra, which makes fertilizer from Indonesia's natural gas to increase rice production.

million of its own foreign exchange, demonstrating the importance of its increased oil revenues. At the same time, the warning of how quickly nature can wipe out hard-won gains when the nation is trying to reach the goal of food self-sufficiency did not go unheeded. The government is not only intensifying efforts to increase rice production but also encouraging production of other food crops. It has a new sense of urgency because of Indonesia's increasing population and the growing world food shortages that indicate the world may be entering a period when a country's ability to buy food may be no guarantee that it can find a seller.

The successful mass guidance system to increase rice cultivation has now been extended to second crops such as soybeans, peanuts, corn, and cassava. There is a major new effort to develop Indonesia's fishing and cattle industries. The Asian Development Bank has loaned the government $2.5 million to establish a fisheries development project in the Riau Islands, off central Sumatra, to be used for purchasing fishing vessels, setting up cold-storage facilities, and acquiring deep-freeze vessels. The World Bank has made a fishing survey off north Sumatra and is considering a project for modernization of the equipment and methods of small fishermen. Cold-storage facilities for fish are being built in central Java. Japanese companies are planning to help Indonesia develop a tuna industry.

An important new mass guidance program for farmers provides for further develpment of "onggole" cows, originally from India, which are good for meat, cross-breeding, and field work. Indonesia is already the biggest cattle-producing country in Southeast Asia, exporting them to Singapore and Hong Kong for reexport to Vietnam and Taiwan and for meat canning. The government hopes to develop an export cattle market in the Phillipines and Malaysia.

By 1974 Indonesia's sugar production is expected to meet all of its domestic requirements and create an export surplus. Many new sugar mills are under construction. In north Sumatra a Swiss company is establishing a 375,000-acre sugar plantation and one of the largest sugar mills in Southeast Asia. This is the first sugar estate on the island, which has excellent sugar-producing potential.

Indonesia is a major exporter of such agricultural products as rubber, copra, tea, coffee, tobacco, palm kernels, palm oil, and, of course, spices. Production of all these products has greatly increased, and many new agricultural enterprises have been launched. The extent of investment by domestic and foreign private enterprises in new agricultural projects is proof of the wisdom of the New Order's dual policy of encouraging local private enterprise and attracting foreign skill and know-how to upgrade Indonesia's agriculture. Since 1968, approved domestic private projects, including agricultural estates, fishing, and husbandry, amount to $155 million of intended investment capital. Foreign private enterprise investments for such projects amounted to more than $106 million by the end of 1972. Many of the projects were joint ventures between foreign investors and Indonesians.

Overall statistics and broad categories do not tell the whole story in a developing country. Imagination and experiments become footnotes in reports even though they may be laying the foundation for important new industries. One such is a joint venture on 25,000 acres in south Sumatra, between Intrada, an Indonesian company, and Mitsubishi Corporation, of Japan, for the cultivation of the quick-yielding crops of castor oil and rosella, a substitute for hemp fiber. Planting of these products was decided upon following two years of research by Mitsubishi. After tests of varieties of castor oil from all over the world, it was finally discovered that two Indonesian varieties were best-suited for production. There is a big market for rosella and castor oil both in Indonesia and in Japan. The project calls for subsequent planting of tobacco and jute and the construction of a jute-bag factory.

Indonesia's greatest success story to date, next to oil, is the rapid development of its timber resources. From only $6 million of timber exports in 1967 there was a dramatic increase to $230 million at the end of 1972, making timber Indonesia's second major export and providing 12 percent of total exports. The value of 1973 timber exports is expected to increase by an additional $100 million. It is the nation's new glamor industry in terms of potential, employment, and establishment of forest-based industries.

The Foreign Capital Investment and Domestic Capital Investment Laws of 1967 are directly responsible for the development of this great natural resource. Prior to them, the cutting of Java's teak forests, a government monopoly carried over from colonial times, accounted for more than half the value of the country's wood exports. Less than a fourth of Java's land area is classified as forest, whereas the wild forests of the larger, undeveloped islands cover 60 percent of Sumatra, 77 percent of Kalimantan, 52 percent of Sulawesi, and 75 percent of Irian Jaya and the Moluccas. All told, forests cover 300 million acres of Indonesia's total land area of 475 million acres. About 60 million acres are suitable for continuous timber production and 45 million for conversion by logging to other forms of agricultural production. Furthermore, Indonesia is situated on the crossroads to timber markets in Japan, Australia, and the Middle East.

As a result of the profit incentives provided by the new investment laws, at the end of 1972 the government had approved a total of approximately $445.4 million of foreign investment in some sixty forestry projects and more than $200 million of domestic investment in about one hundred forty projects. Many of these are joint ventures. About half of the acreage suitable for continuous timber production is under concession. Concession holders pay the government an annual fee of two cents for each acre developed and royalties ranging from fifty cents to seven dollars for every cubic meter of timber, depending on the type. It also receives a 10-percent export tax. Concession rights, usually covering twenty-year periods, may be extended. Future concession agreements may make it compulsory to establish forest-based industries unless they are not economically feasible. This is in line with the government's objectives to reach self-sufficiency in paper products, which a World Bank survey indicates could be achieved by 1975, as well as encourage the establishment of plywood plants for export production and increase sawed exports.

South Korea's success in establishing an immensely profitable plywood export business, even though it has no timber resources, demonstrates how much more Indonesia can do with its abundance of timber. There is a big market in industrialized countries for the expensive tropical hardwood plywood used as decorative paneling, because of the beauty of its natural grain pattern. The plywood grain must be matched visually by workers, which creates an ideal, labor-intensive industry for developing countries. As a developing country, South Korea began its plywood industry by buying logs from the Phillipines. Now it has reached the point where it is a foreign investor. Its first overseas investment, made in 1968, was a $4.5-million lumbering project in a 630,000-acre forest in Kalimantan. Two more Korean lumbering companies have entered into joint ventures with Indonesian partners. These investments have had a striking multiplying effect in showing how a developing country can become a developer. More than a dozen Korean companies have entered into joint ventures in Indonesia in the industrial sector for such projects as construction, a calcium carbide plant, a textile mill, and plants to produce antibiotics, knitted fabrics, soap, cement, monosodium glutamate, and batteries. Korean firms are also interested in investing in a tourist resort, coastal fishing, and corn farming.

Sometimes progress takes an odd turn. Recently a British locomotive firm received an order from the Indonesia Forestry Commission which obliged them to delve into their 1919 archives for the design plans of a steam engine similar to the one they had sent to Java to fill their first export order over a century ago. The Forestry Commission wanted to replace a diesel engine on a narrow-gauge railway to haul logs to sawmills and sawed timber to river outlets. It was an economy move, as there is an ample supply of waste timber to fuel a steam engine but diesel fuel transported a long distance by barge is expensive.

Except for a steam generator for its lighting system, the new locomotive is the same as the 1919 model.

Indonesia's next economic success story—the development of its mineral resources—is well under way with an intended foreign capital investment of more than $1 billion and $85.4 million of domestic capital for exploration and development approved by the government between 1967 and the end of 1972. Indonesia is one of the world's rich mineral areas, as its complex geological history was conducive to high mineralization. However, its complex political history prevented any adequate realization of its potential. Out of the total land area of Indonesia only about 5 percent has been mapped geologically in fair detail, about 75 percent has been only sketchily reconnoitered, while the remaining 20 percent is totally unexplored. Offshore geological studies are just beginning.

Before the new investment laws, Indonesia's principal mineral exports were tin, bauxite, and nickel. Indonesia is in the world's richest tin belt. Tin mining began in 1816, and Indonesia became a major world producer until the Sukarno era when production declined drastically. Under the New Order, the state tin enterprise upgraded existing facilities and doubled tin exports by 1972, making Indonesia currently the world's fourth-largest tin producer.

However, development of new resources was another matter. Like oil, mining requires big-risk exploration capital and large development investments. In addition, a mineral prospect takes much longer to prove and put into production than does an oil prospect. The size and richness of ore deposits must be detailed in order to justify building plants, mills, and facilities. From the beginning of exploration in a favorable area to initial production may take five years or more. The Indonesian government was anxious to generate foreign mining investment, and international mining companies were equally anxious to have an opportunity to participate in what promised to be a new mineral boom.

Three months after the foreign investment law was ratified, Freeport Sulphur, an American mining company, signed the first mining Contract of Work to explore for and mine copper in the rugged Ertsberg mountains of central Irian Jaya, which contain one of the world's great outcrops of base metal. Random samples collected in 1936 indicated it was a high-grade ore deposit, but its remoteness and political events conspired to keep it a geological question mark. In early 1968, after a year of exploration drilling and assays, Freeport announced it had proved 33 million tons of high-grade ore, making Indonesia's first copper mine a major world reserve. In December 1972 the first shipment to Japan, valued at $2.4 million, consisted of 9,000 tons of 30 percent copper concentrates containing small amounts of gold and silver. The development of Freeport's $150-million project, which will produce 65,000 tons of concentrates yearly, is a mining exploration saga. It has opened up Irian Jaya's heart, inhabited up to now only by aboriginal tribes. To construct the mine, tramways to carry ore through the mountains, and the processing mill and port, the company brought in one-thousand nine-hundred Indonesian, American, Korean, Filipino, and Australian workers. Starting with nothing in the middle of nowhere, Freeport has created a thriving, modern outpost of productive civilization.

Freeport, which has the sole responsibility of financing the project, was given a special three-year tax holiday, after which income tax is imposed at 35 percent for seven years and at 41.75 percent for the duration of its thirty-year operating permit. It is also exempted from payment of royalties and land rent.

Contracts of Work with other mining companies also contain tax incentives, with income tax rates that range, according to the mineral being produced, from 35 to 48 percent during the thirty-year operating period that begins after facilities are constructed. However, all other contractors pay land rents and royalties. The amount of royalty depends upon the type of mineral and type of product (raw ore, concentrate, or refined product) and is calculated in relation to the export value of the raw product. Contractors have a three-year exploration period with possible extensions.

The government has opened up fifty-three huge blocks of potentially mineral-bearing land throughout Indonesia for applications from foreign and domestic companies, and by the end of 1972 a total of seventeen contracts were signed in the new mineral hunt that extends throughout the islands. Indonesia has some of the largest known deposits of low-grade nickel in the world, and two major nickel projects are almost ready to go into production. Pacific Nickel Indonesia, a consortium including U.S. Steel and Newmont Mining, is constructing a $500-million mining-through-refining project on Gag Island, near Irian Jaya, which by the late 1970s is expected to produce annually some 100 million pounds of pure nickel worth about $150 million at current market prices. International Nickel Company of Canada is constructing a $135-million first phase of a nickel mining and processing facility in south central Sulawesi, which is expected to begin producing initially, in 1974, some 35 million pounds annually of nickel concentrate worth about $40 million annually. Indonesia Nickel Development, a Japanese firm, made another big nickel discovery in late 1972 in the Halmahera Islands and is preparing for production.

A Dutch company and two Australian companies are exploring for tin, Aluminum Company of America is exploring for bauxite, and a Malaysian company is exploring for granite. General prospecting in widely scattered areas is being carried out by Rio Tinto Zinc, Bethlehem Steel, Kennecott Copper, a Newmont Mining consortium, a Japanese company, and an Australian company. Negotiations are under way with an independent American oil company to explore for uranium.

In mining as well as other industries, foreign companies are under pressure to train and employ Indonesian personnel and to use local products and raw materials. Furthermore, mining contracts provide that foreign companies are obligated to provide a genuine opportunity for Indonesian capital participation as soon as production of the project commences. Just how this will be implemented remains to be seen.

President Suharto constantly reminds the people, on every appropriate occasion, that inviting the participation of foreign

Borobudur, an eighth-century Buddhist temple in central Java, is one of the most magnificent monuments in Asia. Running about three miles around nine terraces, it is a masterwork of stone carving and architectural skill, which was a century in building.

Jakarta, capital of Indonesia, is becoming a skyscraper city under the New Order. This stretch of the main thoroughfare, Djalan Thamrin, is crowned by the heroic statue known as the Welcome Monument, erected by former President Sukarno.

capital investment "does not mean that we are selling ourselves." He points out that in each agreement a term of not more than thirty years is fixed for foreign participation, and in addition not all fields of economic activity are open to foreign capital. "Within the next thirty years it is expected that the advancement of our economy, technology, capital, and business management and the skills of the Indonesian people will be fully capable of any economic activity which at present still needs foreign capital and management," he says. "It should be remembered that economic development means the transformation of economic potentials into real economic strength through capital investment, the application of technology, and the improvement of organizational and managerial ability. Indeed, development should be based on the capability of our own people, but we should take advantage of foreign capital resources, technology, and skill at our disposal, as long as we ourselves lack them and they do not make us dependent forever."

More than 70 percent of all new foreign investment capital committed and spent since the 1967 Foreign Investment Capital Law is for oil, mining, forestry, and agricultural projects. However, companies from twenty-seven countries have invested $696.7 million in textile, chemical, metal products, and other manufacturing plants and $220.1 million in construction, hotels, real estate, business and in other services. The bulk of the capital has come from the United States, Japan, the Phillipines, Canada, Hong Kong, Australia, and Singapore.

Manufacturing accounted for only 9.6 percent of the gross domestic product in 1971, but the textile industry, the most important sector, is growing at a phenomenal rate. It accounts for 40 percent of total foreign manufacturing investment. One of President Suharto's targets from the beginning was to achieve self-sufficiency in clothing. The results to date have far exceeded Five Year Plan expectations. In 1972 Indonesia imported only $19.3 million worth of textiles as contrasted to imports of $58.5 million in 1968. Furthermore, textile experts predict that Indonesia will be the biggest textile producing country in Southeast Asia and an exporting country within the next ten years.

President Suharto is particularly proud of the fact that, aside from the growth and volume of production, quality and designs equal those of foreign textiles. Recently noting the fact that the trademark, *Buatan Indonesia*—Made in Indonesia— had not been put on textile products, he urged domestic and joint foreign venture producers to do so. He wanted the Indonesian public, which is still import-minded, to appreciate the quality of new national textiles.

Although the government provides guarantees, tax holidays, and other incentives for foreign investment, it also imposes limitations and obligations. It does not permit investment in industries vital to national defense such as production of military supplies, nor does it allow foreign investors to have full control of ventures in such fields as aviation, harbors, shipping, railroads, mass media, telecommunications, atomic energy, water supply, and electrical power. There are various ways in which foreign investment in these activities can be made: joint ventures with state enterprises, sale and lease-back of equipment, and profit-sharing management agreements. For example, in 1969 International Telephone and Telegraph built the country's first international satellite communications system, providing telephone, telex, and television channels. The government owns the station but leased it back to ITT for twenty years in return for a share in the net profits.

Foreign investors can have 100 percent equity in high-priority ventures like automobile manufacturing and heavy industry, but the government has a strong policy favoring joint ventures with Indonesian partners. In the early regulations under the investment law, there was a requirement that full tax holidays could not be obtained except by a joint venture if the project involved investment of less than $2.5 million. Because of the shortage of local capital and skills this was not practical, and the government adopted a more flexible policy without a fixed value level. In general, joint ventures have been established on the basis of 30 percent local capital. When a lesser percentage is approved, there is an agreement with the foreign investor to increase local equity within a specified period by sale of shares to Indonesian citizens. The main thrust of government policy is to ensure the participation of its own citizens in new ventures. It is reconciling the rising forces of economic nationalism with the realities and needs of the developing period.

In regard to domestic capital investment, both the government and foreign investors face what President Suharto calls "a very sensitive problem—the relationship between indigenous and nonindigenous enterprises." This stems from the concentration of local economic power in the hands of businessmen of Chinese extraction. It began in colonial days when Chinese were the middlemen between Indonesians and their Dutch rulers and increased under Sukarno's socialism. Indonesian-Chinese businessmen have prospered greatly under the New Order's economic policies, because of both their business experience and their ability to obtain financing from family connections in Singapore and Hong Kong. This has created tensions in the economic community which have become a social and racist problem. President Suharto has appealed to nonindigenous entrepreneurs to seek the participation of indigenous entrepreneurs in their ventures, just as foreign investors are encouraged to seek Indonesian partners. To aid indigenous businessmen, the government has established an investment credit scheme and, under the Domestic Capital Investment Law of 1967, offers them incentives more or less the same as those for foreign investors.

The government also protects and encourages local businessmen by closing a large number of small industries to foreign investment. These involve milk and other dairy products, cigarettes, matches, shoes, paint, nails, flour milling, bicycle tires, tiles, bricks, and a variety of consumer items.

On the other hand, the government has a priority list for foreign investment of industries that are entitled to state bank credit. The list includes agricultural mills, plants, farming, meat and fish canning plants, cold storage, fertilizers, cement, chemicals and pharmaceuticals, textiles, paper and printing plants, metal industries, edible oils, soap, handicrafts, packing and building materials, passenger and goods transport, and hotels outside of Jakarta.

Industrial estates are being built to accommodate domestic and foreign manufacturing firms, providing them with roads, electricity, water, telephone service, and other facilities. The largest one, Pulogadung, which will ultimately cover fifteen hundred acres about thirteen miles east of Jakarta, already houses more than fifty firms manufacturing everything from noodles to sewing machines. When finished, it will have three hundred fifty enterprises employing thirty-five thousand people. Similar estates are in advanced planning stages for Surabaya and Cilacap, both in Java.

A significant new trend is the increasing number of joint ventures between state enterprises and foreign firms. There are about one hundred forty separate central government enterprises engaged in various businesses as a hangover from the

Sukarno era. The New Order is converting as many as possible from national companies, depending upon some government support, to independent, limited-liability companies free to form partnerships with foreign investors. The intent is to make them into profitable enterprises so that their shares can eventually be sold to private Indonesian investors. Foreign firms have already made joint ventures with state enterprises in can making, automotive parts, aluminum extrusions for housing and transport industries, ink, beer, and soft drinks.

On instructions from President Suharto, Pertamina took 60 percent equity in Krakatau, Indonesia's state-owned steel corporation, and is planning a major integrated steel development complex that could eventually cost up to $700 million and produce from 2 to 7 million tons of steel annually. This project was originally begun by Sukarno with Russian aid, but was never completed. Krakatau now has a joint venture with the Hoogoven firm, of the Netherlands, and with International Pipe Industries, of Manila, for first-phase development to produce steel pipe and other steel products required for offshore oil development. Krakatau is also negotiating with Dutch and Australian firms for joint ventures to carry out subsequent phases of the project.

A new kind of joint venture was launched in mid-1973 with the establishment of the Merchant Investment Corporation, Indonesia's first investment finance institution involving private foreign banking and investment houses and Indonesian state commercial banks, whose purpose is to develop a money and capital market in Indonesia. The partners are the state-owned Bank Ekspor-Impor Indonesia, Morgan Guaranty International Finance Corporation, of the United States, the Sumitomo Bank, Ltd., of Japan, and Mees and Hope NV, of the Netherlands. The corporation will provide new sources for medium-term (one to five years) development financing, issue bonds, and invest in new companies until their shares are sold in the market. It will also help foreign and domestic investors find partners for joint ventures and advise on financing, investment, and management. More than twenty five foreign private banking firms are negotiating to set up additional joint venture finance corporations under the same government guidelines. This is another demonstration of foreign confidence in Indonesia's rapidly growing and politically stable market.

Indonesia's sudden growth has been accompanied by growing pains on the part of both investors and government. Antiquated laws, bureaucracy, lack of data for planning, red tape, inadequate infrastructure, energy and water shortages, and a host of other frustrating problems have plagued the establishment of new industries. There was such a backlog of investment applications experiencing long delays in approval that, in mid-1972, President Suharto called for simplification of investment procedures. A major study and recommendations have been completed and improved procedures are being put into effect. The New Order's determination to solve problems, combined with the country's long-range prospects, in great part ameliorate the country's short-term development difficulties.

During the first Five Year Development Plan, the government has concentrated on maintenance, rehabilitiation, and upgrading of existing infrastructure of roads, harbors, ports, airports, and other facilities. The second plan will provide for the expansion of the infrastructure with emphasis on the development of supplies of water and electricity. As this expansion occurs, industrial development will increase more rapidly.

Just as foreign aid was crucial to the New Order's first phase of economic stabilization, it remains crucial during the 1970s for continued development. One of Indonesia's most outstanding achievements is the confidence and enthusiasm it has engendered in foreign governments and world organizations as demonstrated in the amount and quality of their foreign aid and loans. Since Indonesia's major creditors organized the International Governmental Group on Indonesia, or IGGI, to reschedule its debts, Indonesia has received a total of $3,931.9 billion in grants and low-interest, long-term loans from 1966 through fiscal year 1974. Foreign pledges of $724 million in 1972 exceeded the Indonesian government's request by $54 million. Indonesia requested $760 million for 1973/4, and pledges of $876.6 million in part reflect currency devaluations, but some countries increased their aid. The greatest amount of aid has come from the United States, Japan, the International Development Association, the Netherlands, and West Germany. Other participants are the Asian Development Bank, France, Australia, the United Kingdom, Belgium, India, Italy, Canada, Denmark, and New Zealand.

As more projects reach the implementation stage, Indonesia's ability to utilize aid increases. It can absorb greater inputs of capital as the development program gathers momentum. It is proving the truism that success begets success. The government is also proving its ability to meet interest payments and repay its debts. Government economists estimate that such payments, amounting to 20 percent of foreign exchange earnings, do not constitute a heavy burden. In 1972, due to increased exports, installment payments were less than 10 percent of foreign exchange earnings.

In taking inventory in mid-1973, as the nation began the final year of its first Five Year Development Plan, those responsible for the New Order would be justified in taking satisfaction in the battles won in the emergency war for economic and political stability. However, the president and his cabinet are already involved in the most important battle of all—fulfilling the Pancasila principle of social justice.

"Economic development does not automatically bring social justice," President Suharto says, "but it is also clear that without economic development, social justice is just a daydream. Should we exercise social justice under conditions of economic backwardness, then we would only divide poverty. This is obviously not the aim of our struggle. What should be, and has been done, is that right from the start we have laid the basis for social justice in economic development, or at least we are moving in that direction. We are bridging the gap between the haves and the have-nots."

10
Future Harvests

Until recently, in the mountains of central Java some two thousand people living in five adjoining villages stoically accepted as a way of life the necessity to walk two hours each day to obtain water for drinking and cooking. They suffered from skin diseases due to lack of water in which to bathe. Washing clothes was an infrequent luxury. Today, their lives are transformed as the result of the efforts of a group of students and professors from Gadjah Mada University at Jogjakarta.

The students built a system of reservoirs and pipes down the mountainside and across ravines. Pure water from small springs was channeled into large reservoirs and piped into each of the villages. Now there is an abundance of water from the public tap for all the villagers' needs. Outdoor bathhouses and lavoratories have been built. Clothes are washed regularly. Skin diseases are no longer prevalent. The villagers have a new outlook on life and are eager to receive guidance on other projects in which they can cooperate to improve the quality of their living.

This is only one of thousands of small miracles that are occuring throughout Indonesia as the result of a remarkable, innovative melding of the activities of governmental, private, voluntary, and international agencies and organizations. In the absence of adequate government funds to finance social projects on a big scale, the mutual assistance spirit of *gotong-royong* is one of the most dynamic forces at work in Indonesia today. This typically Indonesian approach to solving problems is laying a firm foundation for real social progress when the government can expand its programs, because people at all levels are genuinely involved.

The Gadjah Mada University village water project exemplifies what is happening. The Ministry of Education provides small grants of approximately $2,500 a year to universities for the purpose of "rendering service to society in the solving of its problems." Twenty-five universities have established Bureaus of Community and Public Service. Gadjah Mada University did not have sufficient funds for construction materials to carry out the village water project, so it contacted Australian Aid Abroad which provided the necessary asbestos pipes.

University students are extremely enthusiastic about these programs, as they learn about real community problems while helping solve them. Diponegoro University, on the northern coast of central Java, literally adopted an isolated, poverty-stricken fishing village. The village had no clinic, so, to gain the villagers' confidence, medical students first visited them in their homes, gave them advice, and dispensed medicines. Then the students opened and staffed a clinic. Subsequently, they even introduced a family-planning program. Other students made detailed studies of the villagers' circumstances and needs. Engineering students designed models of inexpensive wood and bamboo houses and showed the villagers how to build them. Agriculture students distributed cocks to improve the quality of chickens and eggs. Economics students worked out a plan to improve their methods of marketing fish and increase their incomes. This pilot experiment developed into a program for other coastal villages.

Other universities throughout Java have extensive action programs in assisting with rice intensification programs, building and stocking fish ponds, introducing new crops and livestock to villagers, and helping them start small village industries.

Students and professors from universities participating in these programs meet annually to exchange information and ideas. As a result of interuniversity cooperation, an Agro-Economic Survey has been established for research projects in agriculture, marketing, and resources development. Numerous workshops have been held with participants from fifteen universities and twenty government offices. The U.S. Agricultural Development Council and the Ford Foundation provide financial assistance. To research policies that may apply to community development work nationwide, a Regional Scientific and Development Center was founded in central Java, with the cooperation of the universities of Gadjah Mada and Diponegoro, various government agencies, and World University Service.

The universities work closely with the government in reporting their findings and making recommendations, which is of great value in helping shape government policies. A new approach to community development is evolving in Indonesia, based on modern techniques and technology, which is unusually sensitive to the cultural and psychological needs of the people as well as their physical ones.

The Ministry of Manpower has inaugurated an effective volunteer work service program for small groups of university graduates to work in villages for a two-year period at a subsistence salary of twelve dollars and fifty cents a month. They are carefully selected and, after a brief training period, are sent to villages to help them modernize at the local level. These trained volunteers have worked with the "mass guidance" agricultural programs, health programs, youth groups, and civic administrations.

The Applied Nutrition Program is a major effort to improve the quality of village life. Initiated by the Ministry of Health, its implementation has been coordinated with the Ministries of

Agriculture, Education, Interior, and Social Affairs and extended to all parts of Java and the outer islands. This program has helped increase the production and consumption of protein foods. The main objectives are to upgrade village poultry flocks and establish fish hatcheries to distribute fingerlings to school and community fish ponds. United Nations agencies–UNICEF, FAO, and WHO–provide financial and technical assistance.

Another strong voluntary movement, which receives assistance and services from government ministries, is Pramuka, the Indonesian scouting movement, of which more than 10 million boys and girls, between the ages of seven and twenty-five are members. Three-fourths of them are in rural areas. Although they follow worldwide scouting principles they are an active part of national community development. The government has given Pramuka an endowment fund, whose interest meets their operational costs. The boy and girl scouts help carry out government programs at the village level. They have been very effective in implementing the Applied Nutrition Program by teaching villagers to milk goats, cows, and water buffalo and to drink the milk. They also teach villagers how to grow soya beans and raise chickens and rabbits. In addition, they help build and maintain dams and roads.

The principle of *gotong-royong* was first institutionalized during the Sukarno era with the formation of Village Social Institutes to provide social welfare services. The New Order has strengthened and expanded this framework, and almost three fourths of Indonesia's fifty-nine thousand villages have been organized. The Social Institutes are voluntary, but government ministries help guide their activities in improving all aspects of village life. The institutes organize cooperatives, classes in home economics and vocational training, preschool programs, adult education classes, youth centers, and clinics and help the needy and underprivileged. They also carry out village projects to build schools and houses and start new agricultural programs.

All the various village programs and projects demonstrate the depth of the people's inner resources, which are only beginning to be developed. President Suharto capitalized on this strength with the annual grants of $300 to each village which are producing such impressive results; the villagers, by contributing their work, increase the value of the grants threefold in village improvements. This is the real base that assures Indonesia of being able to weather the development storms yet to come. It is impossible to feed human attitudes and their multiplying effects into the computers—either the mechanical ones or the human statisticians who analyze statistics without the benefit of understanding what transpires in a simple peasant when for the first time in his life he has enough water, better food, and the knowledge that his children will receive a better education. When he is a participant, he has enlisted as a fighter in a social revolution. At the same time, the awakening hopes and desires which are now universal in Indonesia create greater pressures on the government to fulfill them.

These are the intangible factors that change the course of a nation's history. Furthermore, in regard to Indonesia, one of the facts frequently overlooked by the Western world is that Indonesia is fulfilling a *delayed* revolution. Its leaders today fought for freedom, thought they had achieved it, but were forced to watch the nation become enchained by Sukarno's ruinous socialism and almost be captured by the Communist party. Indonesia was rescued by an anti-Communist, dedicated army, motivated by the same desire for the country's freedom which had been responsible for its original organization to fight the Dutch. The role of the army in Indonesia should not be confused with military seizures of power in other countries. Indonesia's army is committed to the 1945 Constitution and to the goal of achieving Pancasila democracy with social justice for the people. The army has a dominant position in Indonesian affairs, but its role is diminishing as civilian leadership develops.

President Suharto views the army as development forces rather than armed forces. When he was reelected in March 1973 for another five-year term by the People's Consultative Congress, following the popular elections of 1971, it is highly significant that in forming his new cabinet he relinquished his position as defense minister and commander in chief of the armed forces. Furthermore, eighteen of the twenty-two cabinet members are civilians and include the same economic team of advisers that he assembled to guide the nation to recovery in the beginning. The congress also elected another civilian member of his original team as vice-president—Hamengku Buwono IX, the sultan of Jogyakarta.

The World Bank estimates that Indonesia's economic growth rate will average around 7 percent annually over the next five to seven years. However, the government faces two formidable tasks as it begins its new term and the Second Five Year Development Plan of 1974 to 1979. It must accelerate economic development to create jobs for the 1.5 million new workers expected to join the work force every year. It must also allocate its financial and administrative resources so that the fruits of progress are shared more equitably between the rural and urban populations.

The inexorable pressure of population growth is the government's most worrisome problem. Rising unemployment could threaten the government's economic and political stability. Consequently, new economic planning is directed toward labor-intensive, export-oriented industries, rather than technology-intensive industries. The Board of Investment has made a study of the experiences of Malaysia and Singapore in building strong industrial economies on low-cost labor. Indonesian investment priorities will be given to similar labor-intensive industries. With low-cost labor, Hong Kong and Taiwan have also built big export markets. Indonesian wage levels are competitive with those of all these countries.

Development of a skilled labor force is an integral part of the government's job-creation program. The number of technical training centers will be greatly increased, and higher priority will be given to them than to the expansion of universities. Following a three-year study, the Office of Educational Development has worked out a program to overhaul the educational system completely. Since the great majority of students do not go beyond primary school, emphasis will be placed at this level not only on reading, writing, and arithmetic but also on practical subjects related to what the students will be doing when they leave school. Secondary schools also will have vocational emphasis. The new system is designed to produce better farmers, skilled workers, technicians, office workers, and small businessmen. Enrollment in universities will be strictly limited to students with high qualifications. The New Order considers education one of its most serious problems, despite the considerable progress it has made in this field. In 1972, 15.5 million of the 22 million children of primary school age were enrolled, but it is estimated that only 65 percent will finish the sixth grade. Lack of facilities drastically limits the numbers who graduate from secondary schools and high schools. Only some 360,000 go on to universities. Although full implementation of the changes to be made in the educational system will take approximately twelve years, the effects will be felt much sooner, as those who leave school early will be employable and capable of participating in Indonesia's modernization.

Another major solution to the employment problem, as well as a means of increasing foreign exchange earnings, lies in further development of Indonesia's fast-growing tourism. It is a rule of thumb that one tourist dollar spent in a country generates five dollars worth of gross national product, because of the goods and services which must be developed for the tourist trade. Tourist facilities also absorb large quantities of unskilled workers who can be quickly trained to provide service. In 1971, an estimated 180,000 foreign tourists visited Indonesia, twice the total of 1969. The World Bank estimates that during the next ten years the number of long-distance vacationers will grow by as much as 3 million, or approximately double the level of recent years. Indonesia will get a great proportion of this trade as tourists become increasingly aware of the variety and remarkable beauty of its attractions. It is preparing for the expected influx with new luxury hotels and motels already built and under construction in Java, Bali, and Sumatra.

Indonesian handicraft is unique, and the government is assisting the development of handicraft industries for the tourist trade and export. This is another growing employment field. In

Opulently garbed in rich costumes of batik and metallic cloth ornately patterned in gold, these two dancers are part of the color and drama that make up the splendor of traditional Indonesian dances.

1971, the export value of handicrafts, such as woodcarvings, batik cloth, handwoven goods, paintings, and leatherwork doubled in comparison with the previous year.

Indonesia is one of the few Asian countries that still have big undeveloped land frontiers for both agriculture and industry. Most of the new industrial development has taken place in the Jakarta area on Java, where the infrastructure, market, and living conditions are better. However, industrial development is beginning to spread to other parts of Java and the outer islands as construction of power plants, roads, and airports is completed.

The government is accelerating its "transmigration" efforts to encourage families to leave overpopulated Java and become a developing force in the outer islands. In 1972, 11,000 families were transmigrated as compared with 5,000 families the year before. The cost of transmigration is a limiting factor, as it costs $580 to send one person to a resettlement area, including food rations for about twelve months, house, and land. However, when development begins successfully in a new area there is an incentive for people to transmigrate spontaneously. This happened in Sulawesi, where new agricultural development attracted six times as many people from Bali as had relocated in Sulawesi before development began. It is estimated there are 50 million acres of land with good crop potential in Sumatra, Sulawesi, and Kalimantan. The United Nations Development Program and FAO have recently extended $409,000 in aid for aerial photography to provide basic data on the country's general topography, vegetation, and land use to be used in planning further development projects. During the second Five Year Plan, the government target is to increase new land development from the past annual rate of 750,000 acres to 1,250,000 acres a year.

In the long term, family planning is the prime solution to population pressure. In 1970 President Suharto established the National Family Planning Coordination Board, combining government and private activities, which reports directly to him. "We can step up our agricultural, mining, and industrial production, but if these increases are not balanced with controlled population increases, then production output and all the results of our development efforts will be meaningless," he told the people. "We are faced with a compelling, emergency situation."

The population is growing at an annual rate of 2.6 percent. At this rate it will double by the year 2000. Two thirds of the 121 million population are on the islands of Java and Bali, which have a population density of about 1,490 per square mile. The current family-planning program covers these two islands and is based on recommendations of a joint WHO-World Bank mission that visited Indonesia in 1969 at the invitation of the government. During the 1971/72 development there were 500,000 acceptors, which was twice the target that had been set.

In 1972, to meet the nation's emergency and for the first time in the population field, the International Development Association (IDA), an affiliate of the World Bank, and the United Nations Fund for Population Activities (UNFPA) jointly agreed to finance an expanded program. The IDA provided a fifty-year, low-interest credit of $13.2 million and the UNFPA committed an equivalent amount as a grant. The project will support a tenfold increase in nonmedical family-planning field workers who will be available full time to work in villages on a long-range plan. School curricula and other programs have been prepared for national population education. The project is constructing, equipping, and staffing ten schools to train nurse-midwives in six provinces and about two hundred and fifty rural and thirty urban maternal-child health and family-planning centers in east Java. WHO, UNESCO, UNICEF, and the Population Council, of New York, are implementing various parts of the program. The project is planned to slow the growth rate to 2.1 percent, by the prevention of 1.7 million births over the next five years.

The thorniest problem the New Order will tackle in the second Five Year Plan is a more equitable distribution of the benefits of economic progress between regions and classes of people. With an estimated per capita income of about seventy dollars a year in 1971, Indonesia still ranks low in Asia. This is

One of Indonesia's new frontiers is Irian Jaya (West New Guinea), comprising more than one-fifth of the nation's land area and inhabited primarily by members of Stone Age tribes.

about a tenth of Hong Kong's per capita income and half that of the Philippines. However, income per capita does not reflect the reality of what has happened. There has been an increasing migration from the rural areas to the cities seeking jobs that are not available. Consequently, there is a striking contrast between the new wealth that has been generated in the upper and middle classes and the poverty of a great mass of people living at subsistence level. The same uneven distribution has occurred on a regional basis, areas having limited natural resources being bypassed in the rush to develop those with the best resources.

The situation will automatically improve substantially as all the agricultural and industrial expansion projects under way come into production and the government receives increased revenues from oil, mining, and timber. However, the government has set a new, high-priority policy to take concrete action. Specific programs to be financed by increased revenues are being prepared to help raise the standard of living of the have-nots.

The government has been providing aid to regions in the form of annual grants of about twenty-five cents per person to be used for regional self-help projects like those carried out with village grants. The government plans to increase this aid to thirty cents per person in the "minus" regions. This aid will have impact greater than the amount per person indicates, because of the large numbers of people involved, the multiplying effect of their labor contribution and the subsequent improvement of local infrastructure, and increase in agricultural production.

The government is already setting up programs for those people who earn their living at a minimum wage rate, such as fishermen and landless peasants. Cooperatives will be strengthened. Credit systems, together with guidance in production techniques, will be extended on a larger scale to small and medium-size industries and handicraft workers.

In addition to aiding those who have not yet been participating in economic progress, the government in late 1972 redoubled its efforts for efficient tax collection from those who have been prospering and has been taking new measures to dry up sources and opportunities for corruption. Government agencies launched an antismuggling campaign and broke up a major car-smuggling ring. The government's complex classification and rate structure for imports facilitated corruption and rewarded smuggling. Consequently, it adopted the Brussels Tariff Nomenclature for product classification, which simplifies and modifies the old system. In 1973, the third annual increase in government salaries was effected as part of the government's policy of gradually bringing salaries in line with the cost of living and to reduce incentives for corruption. High tariffs have been placed on luxury goods, and President Suharto has sharply criticized wasteful spending and ostentation as being harmful to the country's aspirations of achieving social justice.

During the First Five Year Plan the New Order concentrated on domestic affairs, but foreign affairs are increasing in importance because of the combination of the nation's economic and political stability, its expanded role in international trade, its requirements for international assistance, and changing political events in Asia.

Indonesia is a non-aligned nation, but its political philosophy and economic ties are predominantly oriented to the Western world. Indonesia's stance in its relations with Communist countries is a careful one, displaying no hostility but at the same time being on guard to prevent any opening of relations which could lead to an attempt to revive the Indonesian Communist party. Remnants of the party still exist, and President Suharto warned the nation in 1972 that, although security forces have always been able to detect and destroy Communist cells, it was known that the party plans to try to become active at the beginning of 1975. This is one of the reasons that the government has been extremely cautious and slow in the release of some 100,000 Communists jailed in 1965. By the end of 1971, approximately 70,000 had been released. The majority of the remaining prisoners, who are classified as hard-core activists, are still in jail or in exile on Buru Island, in the remote eastern region. The government is trying to rehabilitate them and release them gradually.

Diplomatic relations between Jakarta and Peking have been frozen since the abortive 1965 coup, because Peking openly supported and financed the Indonesian Communist party as well as helped it plan the coup. When Communist China was accepted as a member of the United Nations in 1971, Indonesia did not oppose its entry, but an ironic twist was that the New Order's foreign minister, Adam Malik, was chairman of the General Assembly.

Visiting Jakarta in 1972, British Foreign Secretary Sir Alec Douglas Home said that it was likely that the rivalry among the major nations for world power and influence would now be decided in Asia. Southeast Asian governments are deeply concerned about the development of their foreign policies to cope with this situation, particularly in view of the withdrawal of U.S. forces from Vietnam and scaling down of its military activities and presence. Indonesian Foreign Minister Malik has warned other Southeast Asian nations against developing a "ghetto mentality" and cautioned them to guard against efforts by the richer nations to use them as tools to wage those nations' wars. "We must recognize the fact that the world had become smaller in terms of time and space, and that countries are becoming more and more interdependent," he says. "This interdependency applies even more to Southeast Asian countries, with our faiths linked closely to each other's well-being, due to the fact that we live on a great crossroad of the world, providing links between two oceans and two continents. Whatever happens on or to these continents, or on and to these two oceans, countries in Southeast Asia are bound to be affected, either favorably or otherwise. It could be that we will be brushed aside by the mighty road-users, or they can adopt another alternative, giving us the task of impartial guardians of the crossroad."

Indonesia is urging the strengthening of the Association of Southeast Asian Nations, ASEAN, of which it is a member along with Malaysia, Singapore, Thailand, and the Philippines. This economic organization is beginning to develop into a regional political alliance. All the member countries are concerned about Maoist underground organizations and Communist infiltrators. ASEAN bloc nations have also indicated that they would welcome the formation of another bloc composed of Burma, Vietnam, Laos, and Cambodia as a buffer to mainland China. Oddly enough, even if North Vietnam should obtain political control of South Vietnam, the ASEAN would not consider the inclusion of Vietnam a disadvantage, as North Vietnam has its own self-interest in not wishing to be dominated by China.

Members of the Southeast Asian bloc are also concerned about Japan's stake in the region. Their raw materials are indispensable to Japan's economy, and there is fear that Japan will become more political, and even militant, in protecting its economic interests. Japan took a firm stand against an Indonesian-Malaysian move to declare the Strait of Malacca a territorial waterway. It was also the first "outside" power to show interest in becoming diplomatically involved in the Indochina cease-fire moves in 1972.

Indonesia exports almost half of its raw material products to Japan, and Japanese manufactured goods are more than a third of its imports. Next to Americans, the Japanese are the biggest foreign investors in Indonesia with $344.5 million committed to projects. The Japanese are highly visible everywhere. The biggest sign on the tallest building in Jakarta is "Toyota." Both Japan and Indonesia are keenly aware of the ambivalence of their relationship.

The American Embassy in Jakarta estimates that U.S. private investment expenditures in Indonesia now cumulate to about $1.5 billion. However, Indonesia's greatest importance to the United States is that as a result of the bulk of this investment, Indonesia has emerged as a potential factor in lessening the American energy crisis. Although most of Indonesia's oil currently goes to Japan, as its vast new oil and gas province develops it has the possibility of becoming an important source of supply to the United States. If the United States continues to pursue its present low-key development of its own oil and gas resources, due to political and environmental restrictions, American government and oil industry experts estimate that by 1985 the United States will have to supply 38 percent of its total energy needs by imports of oil and gas. The majority of these imports are estimated to come currently from the politically unstable Middle East, which has 76 percent of proved world oil and gas reserves. Indonesia's development as a major source of oil and gas will help the United States in its efforts to avoid becoming overdependent on the Middle East, with the complications of a potential Arab-Israeli war and the attendant dangers of a confrontation with the Soviets over energy supplies.

It is typical of President Suharto that he equates Indonesia's role in foreign affairs with economics. He considers development success as the greatest guarantee for peace in Southeast Asia. If the "national resilience" that Indonesia has developed can become regional, he believes that the smaller nations can establish a true concept of Southeast Asian neutrality independent of the big powers. President Suharto illustrates his concept of "national resilience" with the fact that Indonesia was strong enough to survive the international monetary crisis of 1971. Furthermore, during the turbulence of the crisis Indonesia's minister of finance, Ali Wardhana, was elected chairman of the Board of Governors of the International Monetary Fund. Also, the governor of Bank Indonesia, Radius Prawiro, who is now the new minister of trade, was elected chairman of the World Bank. Their appointments were a tribute to Indonesia's economic achievements and the experience it has developed by solving its own problems.

Delegations and officials from other Southeast Asian countries are coming to Indonesia to find out what the open secret of its development success is and to study how Indonesians are going about it. All of the Southeast Asian nations who have offshore oil prospects are patterning their oil development policies on Indonesia's successful production-sharing contracts. The latest group to arrive to study both oil and overall development were leading government officials from Burma, which, like Indonesia, tried socialism but found that it doesn't work. Indonesia now finds itself in the position of being a teacher of "how to plant and grow,"according to its ancient rice proverb, not only for its own people but for the peoples of other nations.

Pragmatism is the trademark of the New Order, as well as a traditional Indonesian trait. Indonesian leaders are reluctant to discuss their successes and prefer to dwell on the problems ahead. "We are only cruising with a tailwind, and that tailwind is all the help we have had," says Mohammed Sadli, a member of the original economic team who is the new minister of mines. "It is what happens in the next ten years that will tell the story. We must not talk about success. If we think we have done too well, we will become too complacent."

There is little danger of Indonesia's becoming complacent, and little doubt that it will achieve its goals. The reason it has received so much help is very simple. It is expressed in the old aphorism, "God helps him who helps himself."

President Suharto puts it another way. "A nation engaged in self-development should possess self-reliance," he told the people. "Strengthening one's self-reliance and increasing one's power to be able to stand on one's own feet are prerequisites for the success of development. Changes come only from a change of mental attitude. If progress in the social field, measured in terms of physical substructures, appears slow, we do not feel downhearted, because our mental progress has made such big strides. That is the true capital for development!"

ACKNOWLEDGMENTS AND APOLOGIES

I wish not only to thank all those who have helped me in every way to write this summary of the exciting, constructive things that have been happening in Indonesia since October 1, 1965, but I wish to apologize to all of those participants in the ever-evolving, hard-won success story whose efforts I have not mentioned or have skimmed over too lightly.

A great body of detailed literature will be published eventually about Indonesia's development when all the national and international surveys are completed and as the nation's five-year plans continue. In the meantime, I have attempted to report the main thrust of what is happening at a timely point—President Suharto's 1973 reelection for another five-year term and the beginning of Indonesia's second five-year development plan in 1974. I also hope that this account will help balance what I consider the unjustified gloomy opinion of what is happening in developing countries which prevails in the mainstream of public consciousness in the industrialized countries. As an American, I have been reinspired by Indonesia and have found there the spirit that was responsible for making my own country great. Americans today need to relearn from Indonesia the importance of mental attitudes in the continuing development of our own country.

I wish to thank President Suharto, the various members of his cabinet with whom I have conferred, and General Ibnu Sutowo for their graciousness in helping me understand the dedication of the generation of 1945 and the meaning of Pancasila.

To my many Indonesian and American friends who have helped me in assembling the information presented here, may I say, from the bottom of my heart—*terima kasih*—thank you.

Ruth Sheldon Knowles

New York, New York
July, 1973

Appendixes

Since Indonesia is becoming the most important developing country in the Asian Pacific region for foreign investment, the following data may be helpful to potential investors in showing its recent growth, key economic factors, and the government's priority projects for foreign investment.

Official publications providing a guide to laws and procedures for foreign investors can be obtained by contacting any of the following agencies:

Foreign Investment Board
Jalan Cut Mutiah 2
Jakarta, Indonesia

Foreign Investment Service
Consulate General of Indonesia
5 East 68th Street
New York, N.Y. 10021

Foreign Investment Service
Embassy of the Republic of Indonesia
2020 Massachusetts Avenue NW
Washington, D.C. 20036

Companies interested in obtaining a sample oil Production-Sharing Contract should contact either of the following agencies:

Oil and Gas Committee
Medan Merdeka Selatan 18
Jakarta, Indonesia

Pertamina
866 United Nations Plaza
New York, N.Y. 10017

INDONESIA'S KEY ECONOMIC INDICATORS, 1967–72

	1967	*1968*	*1969*	*1970*	*1971*	*1972*
GDP[a] at current prices (rupiahs billions)	847.8	1,993.9	2,593.9	3,196.2	3,544.1	4,187.0[b]
% change in real GDP over previous year	–	6.6	7.1	6.9	7.0	–
Population (millions)	110.57	112.83	114.48	116.82	119.20	123.115
GDP per capita, current prices (rupiahs)	7,667.5	17,671.7	22,658.1	27,360.0	29,732.4	–
Exports, fob ($ millions)	665.4	730.7	853.7	1,160.6	1,246.8	1,878.0[b]
Imports, cif ($ millions)	649.2	715.8	780.7	1,001.5	1,173.9	1,619.0[b]
Balance of trade	+16.2	+14.9	+73.0	+159.1	+72.9	+259.0[b]
Jakarta cost-of-living index (1966=100)	206	463	545	613	638	807
Exchange rate (rupiahs per $1)	235	326	326	378	415	415

Money supply (rupiahs millions)	51.5	113.9	180.0	241.1	312.6	468.5
Currency in circulation	34.1	74.7	114.3	152.8	195.5	270.5
Demand deposits	17.4	39.2	65.7	88.3	117.1	198.0
Bank Indonesia assets (rupiahs billions)						
Total assets	151.0	268.0	417.2	526.7	598.6	784.2
Foreign exchange assets	82.6	105.3	64.1	78.3	77.0	245.7
Gold coin and bullion	1.1	1.1	1.6	4.9	5.6	6.9
National budget (rupiahs billions)[c]						
Receipts (actual)	84.9	185.3	334.8	465.1	559.1	751.6
Expenditures (actual)	87.6	185.3	334.7	457.9	540.6	751.6
Balance	-2.7	–	+0.1	+7.2	+18.5	–

SOURCES: Bank Indonesia, Central Bureau of Statistics, and other Indonesian authorities.

NOTES: a GDP = gross domestic product.
b Estimated.
c Fiscal years, beginning April 1.

INDONESIA'S MAJOR NON-OIL EXPORTS, 1968–72
($ millions)

	1968	*1969*	*1970*	*1971*	*1972*
Rubber	176	225	247	199	195
Timber	13	26	101	169	230
Tin	49	53	62	64	67
Coffee	44	51	66	55	72
Palm oil and kernels	26	26	42	52	46
Tea	17	10	18	29	32
Pepper	14	10	3	25	21
Copra and copra cakes	43	21	35	26	17
Tobacco	22	14	12	20	30

SOURCES: Bank Indonesia and other Indonesian authorities.

INDONESIA'S MAJOR IMPORTS BY COMMODITY GROUPS, 1968–72
($ millions)

	1968	*1969*	*1970*	*1971*	*1972*
Capital goods	197.5	238.9	375.5	498.4	411.5
Consumer goods	266.5	221.0	249.5	221.2	295.9[a]
Materials	251.8	320.8	376.5	454.3	579.1[a]

SOURCES: Bank Indonesia and other Indonesian authorities.

NOTE: [a]Estimated.

GROWTH OF INDONESIA'S OIL INDUSTRY

The impressive growth of Indonesia's oil industry can be seen in the statistics covering production and exports, the industry's significance for the Indonesian economy, refinery capacity, oil and gas discoveries, and production-sharing contracts. (Sources are Pertamina and Bank Indonesia.)

Production and Exports

Year	*Crude-Oil Production (barrels per day)*	*Crude-Oil Exports*	*Refined Products Exports*
1966	469.000	260,000	87,000
1967	510,000	309,000	80,000
1968	600,000	398,000	75,000
1969	745,000	517,000	93,000
1970	853,600	625,000	99,000
1971	892,100	656,000	92,000
1972	1,071,700	819,000	126,000
1973[a]	1,300,000	–	–

NOTES: Indonesia's crude oil production is estimated to reach the level of approximately 2,500,000 barrels per day in 1976 and 2,870,000 barrels per day in 1979, not taking into account new discoveries.

[a]January–May

Significance For The Economy

	1968	*1969*	*1970*	*1971*	*1972*
Oil exports as % of total exports	40.7	44.8	38.7	45.2	53.4
State revenues from oil as % of total government revenues	22.2	27.0	28.8	32.9	38.8
Value of crude-oil and refined products exports ($ millions)	297.7	374.6	434.8	594.8	966.9

NOTE: Contributing approximately a two-thirds increase in production value, tax revenues, and foreign exchange earnings, in 1972 the petroleum sector grew perhaps five times more rapidly than the economy as a whole. In 1972 oil production accounted for about 5 percent of total gross domestic product. More importantly, expenditure on oil exploration and development, a barometer of future growth, is expected to grow by another two-thirds in 1973, from $271 million in 1972 to a projected $455 million in 1973.

Refinery Capacity (Barrels Per Day)

1969	261,000
1972/73	411,000
1976/77 (projected)	511,000

New Oil and Gas Discoveries, 1966–mid-1973

Onshore	35
Offshore	21
Total	56

Production-Sharing Contracts, 1966–1972

Total number of contracts:	42
Total square miles, offshore and onshore:	965,000
Total minimum exploration commitment as of December 31, 1972:	$78,200,000
Total spent exploration and development, 1969 through 1972:	$671,900,000

OIL PRODUCTION-SHARING CONTRACTS WITH PERTAMINA AS OF JUNE 30, 1973

		Square Kilometers (approx.)		
Operator	*Area*	*Onshore*	*Onshore and Offshore*	*Offshore*
AGIP	Irian Jaya	–		38,000
	South China Sea	–		43,150
Asamera	North Sumatra	1,150		–
Asia Oil Exploration	West Sumatra	–		27,985
British Petroleum	Makassar Strait	–		9,230
Calasiatic-Topco	Flores Sea	–		27,690
Caltex	Central Sumatra	19,480		–
Cities Service	Java Sea	–		57,700
Cie Française des Pétroles	South Sumatra	6,000		2,300
Continental Oil	South China Sea	–		41,200
Dearborn Computers	Halmahera Islands	–	18,431	–
Frontier	South China Sea	–		42,300
Gulf Oil	South China Sea	–		65,000
	Celebes, offshore	–		40,000
Gulf & Western, Inc.	Ceram area	50,000		–
Hartog	Arafura Sea	–		28,600
Huffco	South Sumatra	1,540		–
	Kalimantan	3,270		–
International Oil	Timor Island	8,125		1,925
IIAPCO	Java Sea	–		42,300
Japex	Makassar Strait	–		8,760
	Malacca Strait	–		8,840
Java Sea Oil Company	Java Sea	–		21,200
Jenney Indonesia Joint Venture	West Sumatra	–		28,330
	East Sumatra	–		
	Simeuloe Island	1,800		1,800

Kyushu Oil	Java Sea	–		49,800
	Makassar Strait	–		25,390
Mobil (Asamera)	North Sumatra	1,640		–
Mobil (Refining Associates)	North Sumatra	750		–
Mobil	Malacca Strait	–		15,960
	North Sumatra	–	?	–
North Sumatra Oil (Nosodeco)	North Sumatra	1,390		–
Phillips	Irian Jaya	–	93,860	–
Refining Associates	North Sumatra	13		–
Shell	East Kalimantan	12,500		–
Sinclair	Java Sea	–		21,150
Stanvac	Central Sumatra, South Sumatra	4,450		–
Union Oil	East Kalimantan	1,540		3,300
	North Sumatra	38,400		–
	Central Sumatra	2,120		–
Wendell Phillips	Irian Jaya	12,352		–

FOREIGN AID AND LOAN ACCOUNTS, INTERGOVERNMENTAL GROUP ON INDONESIA

Country	*Amount ($ millions)*
United States	819.5
Japan	505.1
International Development Association	292.9
Netherlands	283.7
West Germany	172.1
Asian Development Bank	61.5
France	58.1
Australia	55.0
United Kingdom	44.7
Belgium	14.6
India	13.7
Italy	n.a.
Canada	4.9
Denmark (nonmember)	4.2
New Zealand	1.3
Total, July 1, 1966–July 30, 1972	2,331.3
Fiscal year 1972/73 total	724.0
Fiscal year 1973/74 total	876.6
Total, July 1, 1966–July 30, 1974	3,931.9

NOTE: The growth and stability of the Indonesian economy has continued to generate foreign confidence in Indonesia's economic prospects. Foreign pledges of $724 million in 1972 exceeded by $54 million the Indonesian government's request. Indonesia requested $760 million for 1973/74 and pledges of $876.6 million reflected currency devaluations (in part), but some countries increased their aid. Utilization of foreign assistance is increasing as more projects reach the implementation stage. As Indonesia's development program gathers momentum it has the ability to absorb greater inputs of capital.

SOURCES: Indonesian government and other Indonesian authorities.

SELECTED LIST OF COMPANIES INVESTING IN MINING, AGRICULTURAL, AND INDUSTRIAL PROJECTS IN INDONESIA 1966–1973

Freeport Sulphur
Pacific Nickel
Alcoa
Bethlehem Steel
U.S. Steel
Newmont Mining
International Nickel of Canada
Kennecott Copper
Indonesia Nickel of Japan
Weyerhaeuser Company
Ford
Phillip Morris
Dillingham Corporation
Goodyear
Uniroyal
Max Factor
Chesebrough-Pond
Abbott Laboratories
Pfizer
Richardson-Merrell
Squibb
C. Itoh & Company
Mitsui & Company
Marubem Corporation
Toray Industries
Asohi Glass
Toyoto
Kawasaki
Matsushita Electric Company
Philips
Hoogovens
Ciba-Geigy
Bayer
Farbwerke Hoechst AG
British Tobacco Company
Rheem Australia
Kiwi International
Australian Consolidated Industries
Australian Dairy Produce Board
Century Storage Battery
Comaico

PRIORITY PROJECTS FOR FOREIGN INVESTMENT IN INDONESIA AS OF DECEMBER 1972

1. Projects in agriculture, plantations, and husbandry.
   a) Projects that directly support the agricultural sector (e.g., plantations, husbandry, pesticide spraying).
   b) Industries that support agriculture (e.g., agricultural equipment, fertilizers).
   c) Industries that process agricultural products (e.g., rice mills, crumb-rubber factories).

d) Processing of forest products (e.g., sawmills, pulp plants, paper mills).
e) Fishery projects with integrated processing facilities (e.g., cold storage, canning factories).

2. Industries whose raw materials are 50 percent derived from domestic sources.

3. Industries that produce raw materials and auxiliary goods for other industries.

4. Basic industries (e.g., iron and steel industries, shipbuilding).

5. Infrastructure projects (roads, ports, bridges, telecommunication channels, irrigation, electricity).

6. Fully integrated textile industries.

7. Industries that produce rice, soap, cooking oil, sun-dried fish, kerosene, batik, cotton textiles, salt, and sugar.

8. Industries that produce 50 percent or more of their products for export.

9. Industries that produce chemical fibers.

10. Chemical industries.

11. Industries that manufacture medicines and health equipment.

12. Tourism projects (with priority to hotels).

13. Transportation.

14. Other ventures that, considering their urgency, are considered of high priority in accordance with the following criteria:
a) Amount of invested capital;
b) Capacity to absorb labor;
c) Location of the undertaking outside Jakarta/Java.

SOURCE: Foreign Investment Board.

INDONESIA: FOREIGN INVESTMENT (excluding oil) 1967-1972
Intended Capital of Approved Projects (in U.S. millions of dollars

(According to sector)

	1967	1968	1969	1970	1971	1972	Total
Agriculture	7.7	22.3	2.5	60.5	6.8	6.5	106.3
Forestry	5.6	67.8	286.5	32.8	12.0	40.7	445.4
Mining	140.0	157.0	326.7	75.0	79.0	225.0	1002.7
Manufacturing							
–textile	–	1.0	36.6	38.2	107.2	96.7	279.7
–chemical	15.7	15.2	14.3	32.6	19.9	14.3	112.0
–metal products	1.6	32.2	7.6	35.7	13.0	16.0	106.1
–other	13.7	21.4	16.6	52.0	64.8	30.4	198.9
Construction and Hotels	–	7.8	26.7	9.4	50.8	40.1	134.8
Real estate and business services	2.0	2.6	1.3	10.4	3.5	24.8	44.6
Other services	6.1	3.7	10.5	6.2	10.2	4.0	40.7
	192.4	331.0	729.3	352.8	367.2	498.5	2471.2

According to Country

	1967	1968	1969	1970	1971	1972	Total
U.S.A.	162.2	17.0	192.8	71.5	19.5	27.2	490.2
Japan	3.2	14.9	114.9	36.8	107.9	65.8	344.5
Philippines	3.0	5.0	251.1	3.0	1.9	7.2	271.2
Canada	–	151.1	–	–	1.2	–	152.3
Hong Kong	–	12.3	15.1	16.8	50.6	45.0	139.8
Australia	3.1	0.1	0.9	5.6	81.3	2.2	93.2
Singapore	–	4.9	14.8	12.1	15.9	21.7	69.4
South Korea	–	48.0	–	3.4	–	2.4	53.8
Netherlands	0.3	23.1	0.7	10.4	2.4	7.7	44.6
U.K.	0.6	0.3	3.0	21.6	4.1	1.2	30.8
Switzerland	–	3.3	0.3	13.2	–	3.5	20.3
West Germany	0.2	5.4	2.5	8.4	2.6	0.9	20.0
Combined	4.0	0.3	77.4	88.1	16.8	240.1	426.7
Other countries (15)	10.9	19.6	21.6	23.2	23.4	19.4	118.1
Indonesian partner capital	4.9	25.7	34.2	38.7	39.6	53.2	196.3
	192.4	331.0	729.3	352.8	367.2	498.5	2471.2

Total number of projects approved by the end of 1972: 549

Source: Foreign Investment Board

Bibliography

As this book is primarily concerned with Indonesia's economic development since 1965, the major part of printed source material for my research is contained in innumerable reports, reviews, and documents prepared by departments and agencies of the Indonesian government, the United States government, the United Nations, private banks, and economic and business journals.

In regard to available, pertinent historical, economic, and general books for anyone who wishes to read further, the ones I found most rewarding and interesting are the following.

Adams, Cindy. *Sukarno, An Autobiography.* Bobbs-Merrill, New York, 1965.

Bartlett, Barton, Bartlett, Fowler, Hays. *Pertamina: Indonesian National Oil.* Amerasian Ltd., Singapore, 1972.

Baum, Vicki. *Tale of Bali.* Literary Guild, New York, 1937.

Business International Asia/Pacific Ltd. *Indonesia: Business Dynamics of a New Frontier.* Business International Asia/Pacific Ltd., Hong Kong, 1973.

Business International Corporation. *Doing Business In The New Indonesia.* Business International Corporation, New York, 1968.

Geertz, Hildred. *Indonesian Cultures and Communities.* HRAF Press, New Haven, 1963.

Hanna, Willard. *Eight Nation Makers–Southeast Asia's Charismatic Leaders.* St. Martin's Press, New York City, 1964.

Jones, Howard Palfrey. *Indonesia: The Possible Dream.* Harcourt, Brace, Jovanovich, New York, 1971.

K'tut Tantri. *Revolt In Paradise.* Harper and Row, New York, 1960.

Roeder, O.G. *The Smiling General.* Gunung Agung, Jakarta, 1970.

Vlekke, Bernard H.M. *Nusantara: A History of Indonesia.* W. van Hoeve Ltd., The Hague, 1965.

Index

MOBIL
ASAMERA
PERTAMINA-NOSODECO
JAPEX
Ind. Gulf Oil
MOBIL
REFICAN
Ashland
JENNEY
SUMATRA
UNION
MOBIL
CALASIATIC/
TOPCO
UNION
CALTEX
CALASIATIC/
TOPCO
KONDUR
Arco
STANVAC
TOTAL
CFP
TOTAL
PERTAMINA
Stanvac
JENNEY
Marathon
VICO/
HUFFINGTON
PERTAMINA
1100 miles
Palembang
STANVAC
PERTAMINA
JENNEY
WHITESHIELD
Amin Oil
IIAPCO
JAKARTA
IIAPCO
Arco
PERTAMINA
JAVA
LEMIGAS
Surabaya
PERTAMINA
JAVA SHELL
SAMUDERA INDONESIA
AGIP
AGIP
CONTINENTAL
CONTINENTAL
KALIMANTAN
PERTAMINA
VICO/HUFFINGTON
PERTAMINA
Pexamin
CONTINENTAL
UNION
Bandyarmas
KYUSHU
Ashland
AUSTRALIA DRILLING CO.
Cities Service
BALI